Paul Matt's Scale Airplane Drawings
Volume 2

Drawn by the late, great

Paul Matt

Published as part of the
"Aviation Heritage Library Series"
by

SunShine House, Inc.
P.O. Box 2065
Terre Haute, IN 47802
© Copyright Historical Aviation Album
SunShine House, Inc.

(812) 232-3076

Publishers: Alan Abel, Drina Welch Abel

First Printing, 1991
Second Printing, 1992
Printed in the United States of America
ISBN 0-943691-05-2

The Aviation Heritage Library Series is published to preserve the history of the men and women, and of their airplanes, during the era of the Golden Years of Aviation. The books in the series include:

The Welch Airplane Story by Drina Welch Abel
It's a Funk! by G. Dale Beach
The Luscombe Story by John C. Swick
The Earhart Disappearance - The British Connection by James A. Donahue
Ryan Sport Trainer by Dorr Carpenter
Aeronca — Best of Paul Matt
Ryan Broughams and Their Builders by William Wagner
The Corsair and other— Aeroplanes Vought by Gerald P. Moran
Peanut Power by Bill Hannan
Paul Matt Scale Airplane Drawings, Vol.1
Paul Matt Scale Airplane Drawings, Vol.2
Roosevelt Field — World's Premier Airport by Joshua Stoff & William Camp
WACO — Symbol of Courage & Excellence by Fred O. Kobernuss
The Taylorcraft Story by Chet Peek
Visions of Luscombe — The Early Years by James B. Zazas

The publisher of this book calls specific attention to the fact that the plans, specifications and drawings which have been reproduced and presented in this book cannot be relied upon as a basis for reproducing any part or portion of an airplane which is the subject matter of this publication.

The publisher of this book makes no representation or guarantee regarding the engineering accuracy of the data reflected and shown in the plans, specifications and drawings reproduced and presented in this book and do expressly disclaim any responsibility for any part or portion of an airplane reproduced therefrom.

First Edition
Printed in the United States of America

ISBN 0-943691-05-2

3-View Scale Drawings by Paul Matt

Volume 1 contains all airplanes listed from A through G. Volume 2 contains all airplanes listed from H through W. If you would like individual, large sheets of any of the below listed scale drawings, send request, catalog #, appropriate fee, and shipping charge of $2.50. If you want your drawing rolled in a tube – add $4.00. Clear, concise, uncluttered 3- and 4-view engineering presentations, in large (17" x 22"), easy to work with standard scales. Drawing is highly detailed and is an accurate working print with all dimensions, specifications, airfoils, cross sections, templates, and color scheme included. These drawings are part of the collection of the internationally acclaimed and highly honored *Historical Aviation Album*. All orders are folded unless a tube is ordered. Each has 2 to 6 large sheets. **Rolled in tube– add $4.00.**

DRAWING SCALE:
CODE: Last letter of catalog numer indicates scale of drawing.
A = 1:24 .... 1/2" = 1'
B = 1:16 .... 3/4" = 1'
C = 1:48 .... 1/4" = 1'
D = 1:32 .... 3/8" = 1'
E = 1:8 ... 1 1/2" = 1'
F = 1:6 .... 2" = 1'

Volume 1 contents:

_Aeromarine 39B	(#9-78A, 3 shts.)	$9.50
_Aeronca 7AC "Champion"	(#AM-41A, 2 shts.)	8.00
_Aeronca C-2	(#10-80A, 2 shts.)	8.00
_Aeronca C-3 "Collegian"	(#10-81A, 2 shts.)	8.00
_Aeronca C-3 "Master"	(#10-82A, 2 shts.)	8.00
_Aeronca K	(#15-103A, 2 shts.)	8.00
_Aeronca LB "Low wing"	(#AM-60A, 2 shts.)	8.00
_Alcor C.6.1 Jr. Transport	(#4-33D, 2 shts.)	8.00
_Anderson-Greenwood AG-14	(#12-94A, 2 shts.)	8.00
_Beechcraft D-18S Twin	(#6-59A, 3 shts.)	9.50
_Bell P-39Q "Airacobra"	(#1-7A, 2 shts.)	8.00
_Berckmans "Speed Scout"	(#2-12B, 2 shts.)	8.00
_Berliner-Joyce OJ-2	(#AM-48A, 2 shts.)	8.00
_Berliner-Joyce XF3J-1	(#15-104A, 2 shts.)	8.00
_Boeing 307 "Stratoliner"	(#7-63C, 2 shts.)	8.00
_Boeing F3B-1	(#AM-56A, 2 shts.)	8.00
_Boeing XF7B-1	(#AM-38A, 2 shts.)	8.00
_Brewster F2A-3 "Buffalo"	(#5-47A, 2 shts.)	8.00
_Cessna 120/140	(#1-9A, 2 shts.)	8.00
_Cessna C-37 "Airmaster"	(#6-50A, 2 shts.)	8.00
_Cessna T-50 "Bobcat"	(#17-115A, 2 shts.)	8.00
_Consolidated P2Y-2	(#9-71C, 3 shts.)	9.50
_Consolidated PBY-5A "Catalina"	(#17-114C, 4shts.)	11.00
_Curtiss "Carrier Pigeon I"	(#2-15A, 2 shts.)	8.00
_Curtiss 1st Milit. Tract. S.C. No. 21/22 1913.	(#6-55A, 2 shts.)	8.00
_Curtiss A-3B "Falcon"	(#AM-39A, 2 shts.)	8.00
_Curtiss AT-9 "Jeep"	(#2-18A, 2 shts.)	8.00
_Curtiss B-2 Condor Bomber	(#18-116C, 2 shts.)	8.00
_Curtiss B-20 Condor Transport	(#18-117C, 2 shts.)	8.00
_Curtiss F Boat	(#1-1A, 2 shts.)	8.00
_Curtiss F92 "Sparrowhawk"	(#18-118A, 2 shts.)	8.00
_Curtiss MF-K-6 "Seagull"	(#AM-37A, 2 shts.)	8.00
_Curtiss P-36 "Hawk"	(#7-62A, 2 shts.)	8.00
_Curtiss P-6E "Hawk"	(#5-42A, 2 shts.)	8.00
_Curtiss PW-8	(#10-79A, 2 shts.)	8.00
_Curtiss R-6 Racer	(#HS-73A, 2 shts.)	8.00
_Curtiss SC-1 "Seahawk"	(#1-8A, 2 shts.)	8.00
_Curtiss SNC-1 "Falcon"	(#AA-65A, 2 shts.)	8.00
_Curtiss SO3C-1 "Seagull"	(#AM-34A, 2 shts.)	8.00
_Curtiss Twin JN	(#3-21A, 2 shts.)	8.00
_Curtiss-Cox "Texas Wildcat"	(#2-14B, 2 shts.)	8.00
_Curtiss-Wright "Condor II"	(#1-6C, 2 shts)	8.00
_Curtiss-Wright CW-1 "Jr."	(#11-88A, 3 shts.)	9.50
_Douglas 0-2	(#11-86A, 3 shts.)	9.50
_Douglas 0-2H	(#11-87A, 2 shts.)	8.00
_Douglas 0-38	(#12-92A, 2 shts.)	8.00
_Douglas 0-38E	(#12-93A, 2 shts.)	8.00
_Douglas M-2	(#14-68A, 2 shts.)	8.00
_Douglas A-20G "Havoc"	(#15-102D, 6 shts.)	13.00
_Etrich 1913 Taube	(#TB-105D, 3 shts.)	9.50
_Fairchild FC-1	(#17-111A, 2 shts.)	8.00
_Fairchild FC-2	(#17-112A, 2 shts.)	8.00
_Fairchild FC-2W "Stars & Stripes"	(#17-113A, 2 shts.)	8.00
_Fairchild M-62, PT-19 "Cornell"	(#AA-69A, 2 shts.)	8.00
_Fokker T.5 Netherlands Bomber	(#AA-61C, 2 shts.)	8.00
_Gallaudet D-1	(#3-20A, 2 shts.)	8.00
_General Aviation Clark GA-43	(#13-95D, 2 shts.)	8.00
_Grumman F-11F-1 "Tiger"	(#2-19D, 2 shts.)	8.00
_Grumman FF-1	(#AA-77A, 2 shts.)	8.00
_Grumman G-44 "Widgeon"	(#AM5-30A, 2 shts.)	8.00
_Grumman J2F-5 "Duck"	(#6-53A, 3 shts.)	9.50

Volume 2 contents:

_Heath LNB-4 Parasol	(#4-35B, 2 shts.)	$8.00
_Howard DGA-15P	(#AM-52A, 2 shts.)	8.00
_Howard DGA-3 "Pete"	(#12-89A, 2 shts.)	8.00
_Howard DGA-4 "Mike"	(#13-90A, 2 shts.)	8.00
_Howard DGA-5 "Ike"	(#13-91A, 2 shts.)	8.00
_Howard DGA-6 "Mr. Mulligan"	(#14-99A, 3 shts.)	9.50
_Hughes 1B Longwing Racer	(#16-108A, 3 shts.)	9.50
_Hughes 1B Shortwing Racer	(#16-107A, 3 shts.)	9.50
_Laird LC-DW-300 "Solution"	(#7-64A, 2 shts.)	8.00
_Laird LC-DW-500 "Super Solution"	(#8-74A, 2 shts.)	8.00
_Laird-Turner LTR-14	(#10-83A, 2 shts.)	8.00
_Lavochkin LA-7	(#AA-46A, 4 shts.)	11.00
_Lockheed F-80 "Shooting Star"	(#2-10A, 2 shts.)	8.00
_Lockheed Model 9 "Orion"	(#3-22A, 2 shts.)	8.00
_Lockheed PV-1 "Ventura"	(#E-101C, 3 shts.)	9.50
_LWF Cato Model L "Butterfly"	(#10-84A, 2 shts.)	8.00
_LWF Model G-2	(#2-13A, 2 shts.)	8.00
_LWF Model H "Owl"	(#11-85C, 2 shts.)	8.00
_Martin BM-1/2	(#1-5A, 2 shts.)	8.00
_Martin T4M-1, Great Lakes TG-1	(#4-29D, 2 shts.)	8.00
_Martin TT, 1913 Trainer	(#5-40A, 2 shts.)	8.00
_Messerschmitt ME 109E-3	(#AA-75A, 2 shts.)	8.00
_Morehouse 2 Cyl. Aero Engine	(#3-27E, 1 sht.)	8.00
_Navy-Wright NW-1 Mystery Racer	(#4-31A, 2 shts.)	8.00
_Navy-Wright NW-2 Mystery Racer	(#4-32A, 2 shts.)	8.00
_North American 0-47A	(#13-98A, 3 shts.)	9.50
_North American AT-6D	(#16-106A, 3 shts.)	9.50
_North American XB-70-1 "Valkyrie"	(#7-66G, 3 shts.)	9.50
_Packard-Lepere LUSAC-11	(#1-3A, 2 shts.)	8.00
_Pfitzner 1910 Monoplane	(#2-11A, 2 shts.)	8.00
_Piper J-3 "Cub"	(#18-121A, 2 shts.)	8.00
_Piper J-4 "Cub Coupe"	(#18-122A, 2 shts.)	8.00
_Piper PA-12 "Super Cruiser"	(#4-36A, 2 shts.)	8.00
_Republic RC-3 Seabee	(#16-109A, 3 shts.)	9.50
_Rover Inverted Aero Engine	(#5-45F, 1 sht.)	5.00
_Ryan B-5 "Brougham"	(#RB-96A, 2 shts.)	8.00
_Ryan FR-1 "Fireball"	(#3-26A, 2 shts.)	8.00
_Ryan SCW Low Wing	(#AM9-49A, 2 shts.)	8.00
_Ryan ST-A	(#9-76A, 2 shts.)	8.00
_Seversky BT-8	(#3-23A, 2 shts.)	8.00
_Seversky P-35	(#AA-57A, 2 shts.)	8.00
_Sikorsky S-39B	(#14-100A, 5 shts.)	10.50
_Standard J-1	(#17-110A, 2 shts.)	8.00
_Taylor E-2 "Cub"	(#18-119A, 2 shts.)	8.00
_Taylor J-2 "Cub"	(#18-120A, 2 shts.)	8.00
_Thomas-Boeing MB-3A		
_Thomas-Morse MB-3	(#4-28B, 2 shts.)	8.00
_Timm TC-170 "Collegiate"	(#8-72A, s shts.)	8.00
_Verville R-3 Racer	(#6-58A, 2 shts.)	8.00
_Vought F4U-1 "Corsair"	(#AA-51A, 3 shts.)	9.50
_Vought SBU-1	(#3-24A, 2 shts.)	8.00
_Vought XF5U	(#8-70A, 3 shts.)	9.50
_Vultee V-1A	(#2-16A, 2 shts.)	8.00
_Waco UMF/YMF-5	(#13-97A, 3 shts.)	9.50
_Waco UPF-7	(#8-67A, 2 shts.)	8.00
_Waco YKS-6 Cabin	(#2-17A, 2 shts.)	8.00
_Waterman "Arrowbile"	(#3-25A, 2 shts.)	8.00
_Waterman "Gosling" Racer	(#1-4B, 2 shts.)	8.00
_Wright F2W-1 Racer	(#5-43A, 1 sht.)	4.50
_Wright F2W-2 Racer	(#5-44A, 2 shts.)	8.00
_Wright Brothers 1903 Flyer	(#0-123A, 2 shts.)	8.00
_Wright-Martin V	(#1-2A, 2 shts.)	8.00

send to: **Aviation Heritage**
P.O. Box 2065
Terre Haute, IN 47802

DEDICATION

Dedicated to the late
Paul R. Matt
and his wife, the late
Joan Woeste Matt
who both contributed greatly to the
recording of aviation history

Heath LNB-4 Parasol

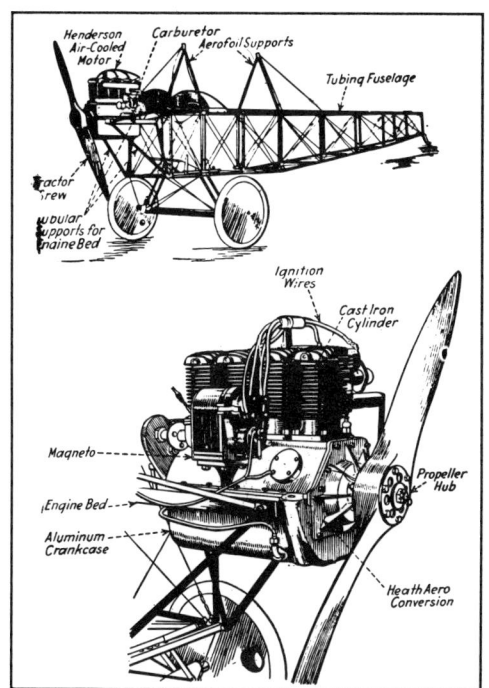

Heath Parasol and Henderson converted motorcycle engine installation.

Heath Parasol on floats. Ed Heath by aircraft.

Heath LNB-4 Parasol.

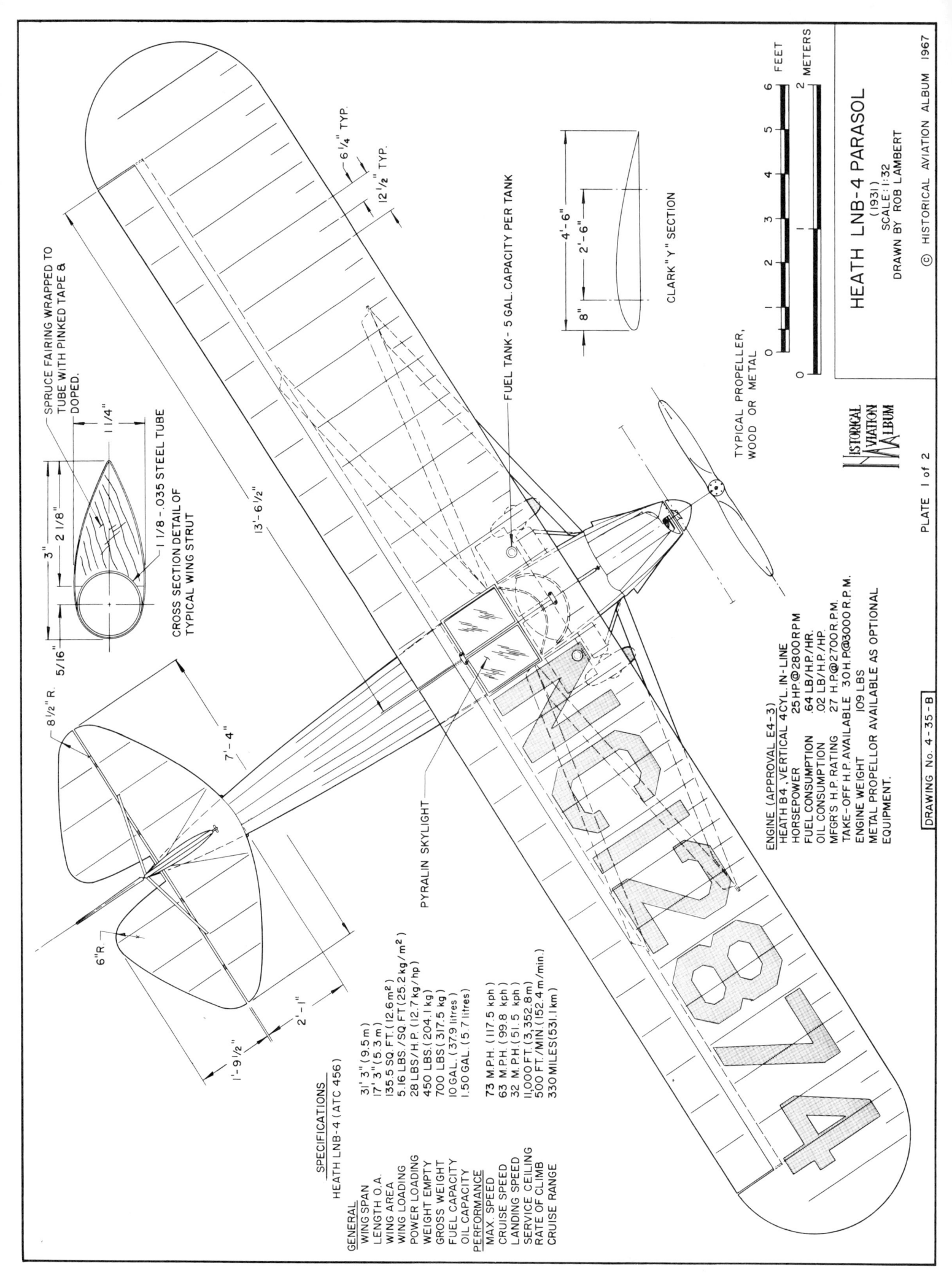

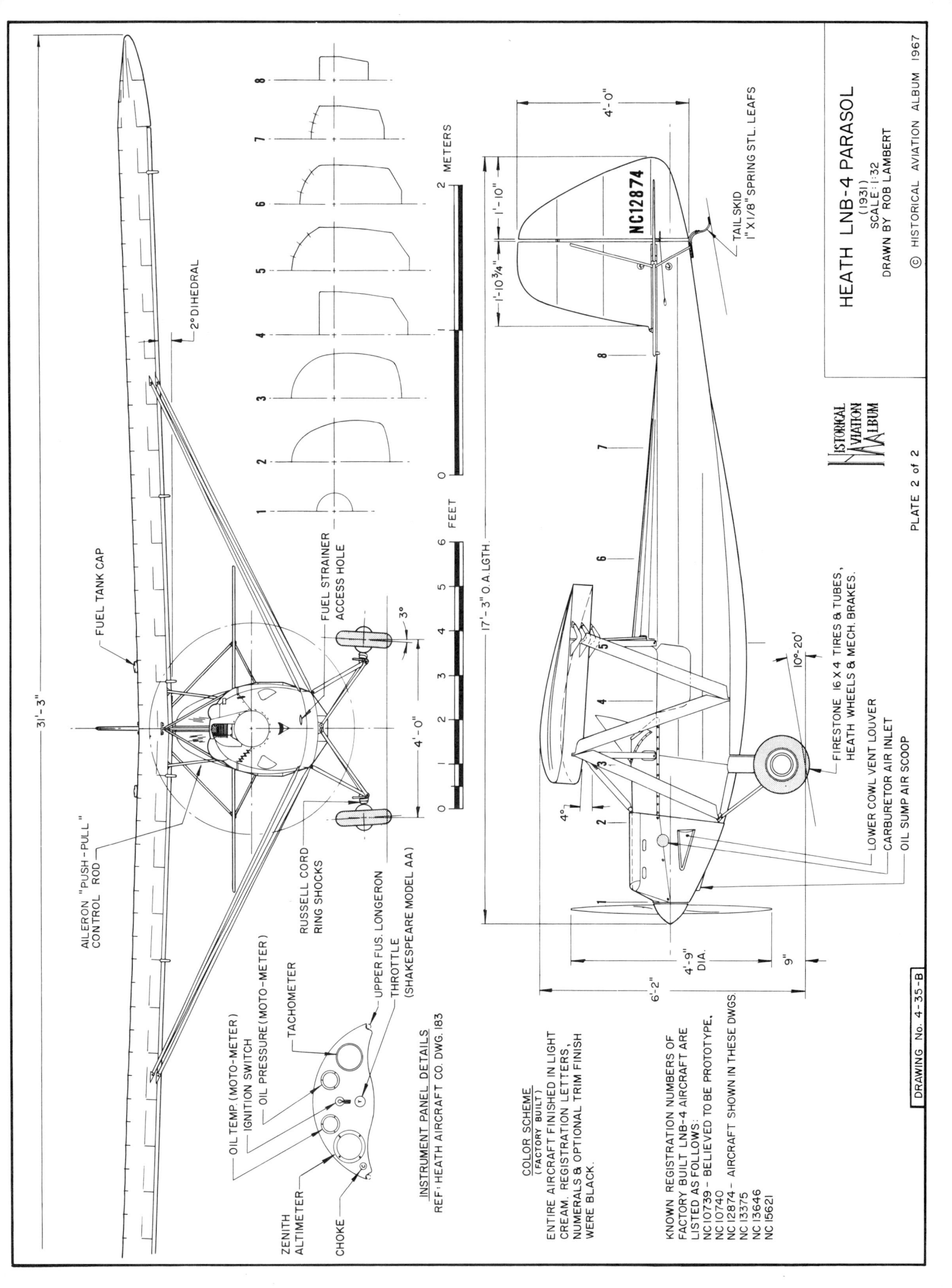

Howard DGA 15P

Dark blue and red DGA-15P of the Aeronautical Commission, State of West Virginia, visits Cincinnati, Ohio – 1956.

Howard DGA-15P Modified post WWII.

Howard DGA-15P.

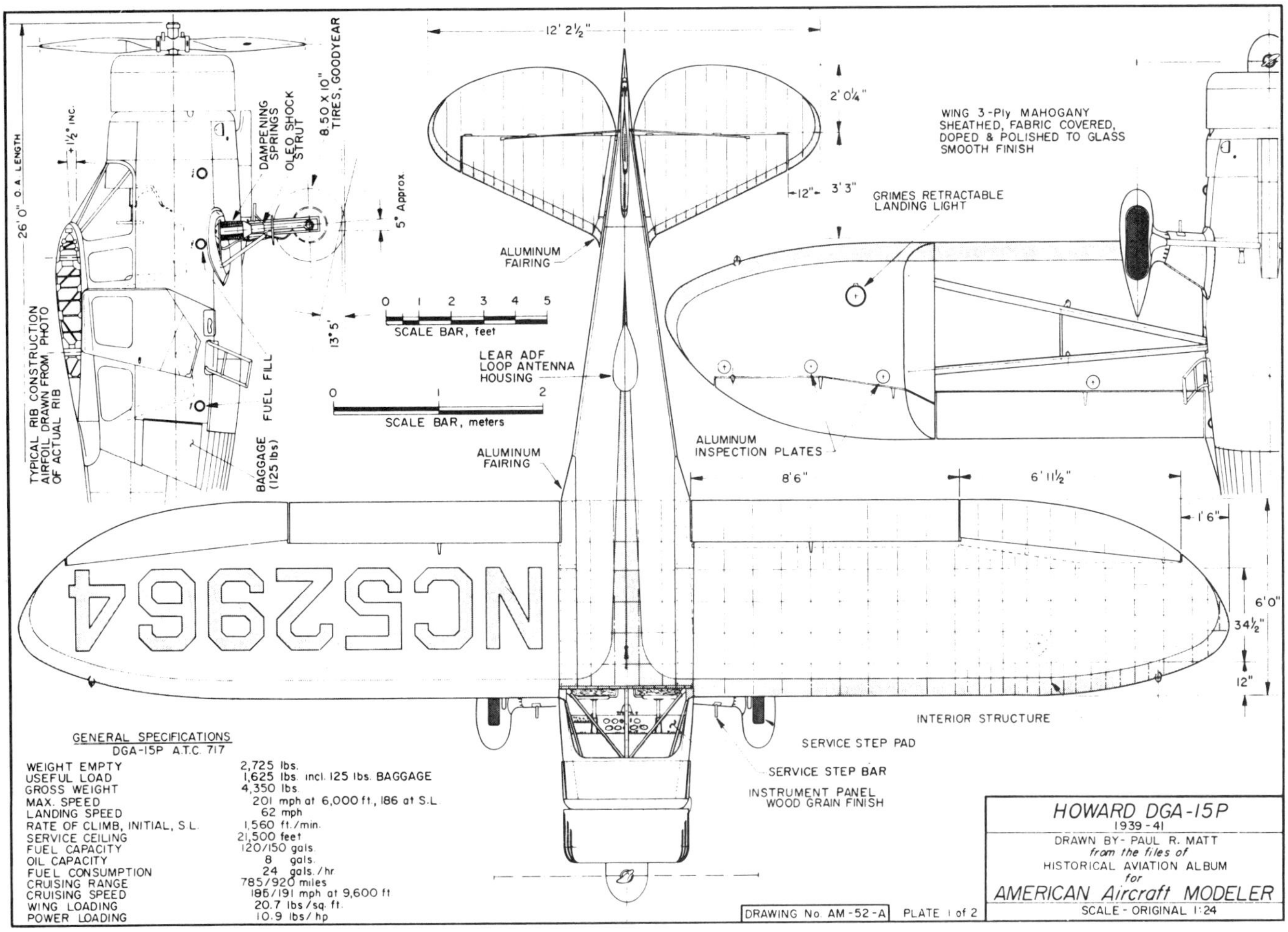
12' 2½"
2' 0¼"
WING 3-Ply MAHOGANY SHEATHED, FABRIC COVERED, DOPED & POLISHED TO GLASS SMOOTH FINISH
12" 3' 3"
GRIMES RETRACTABLE LANDING LIGHT
ALUMINUM FAIRING
0 1 2 3 4 5
SCALE BAR, feet
LEAR ADF LOOP ANTENNA HOUSING
0 1 2
SCALE BAR, meters
ALUMINUM FAIRING
ALUMINUM INSPECTION PLATES
8'6"
6' 11½"
1'6"
6'0"
34½"
12"
NC52964
INTERIOR STRUCTURE
SERVICE STEP PAD
SERVICE STEP BAR
INSTRUMENT PANEL WOOD GRAIN FINISH
26' 0" O.A. LENGTH
1½° INC.
DAMPENING SPRINGS
OLEO SHOCK STRUT
8.50 X 10" TIRES, GOODYEAR
5° Approx
13° 5'
TYPICAL RIB CONSTRUCTION AIRFOIL DRAWN FROM PHOTO OF ACTUAL RIB
FUEL FILL
BAGGAGE (125 lbs)
GENERAL SPECIFICATIONS
DGA-15P A.T.C. 717
WEIGHT EMPTY 2,725 lbs.
USEFUL LOAD 1,625 lbs. incl. 125 lbs. BAGGAGE
GROSS WEIGHT 4,350 lbs.
MAX. SPEED 201 mph at 6,000 ft, 186 at S.L.
LANDING SPEED 62 mph
RATE OF CLIMB, INITIAL, S.L. 1,560 ft./min.
SERVICE CEILING 21,500 feet
FUEL CAPACITY 120/150 gals
OIL CAPACITY 8 gals.
FUEL CONSUMPTION 24 gals./hr
CRUISING RANGE 785/920 miles
CRUISING SPEED 186/191 mph at 9,600 ft
WING LOADING 20.7 lbs/sq. ft.
POWER LOADING 10.9 lbs/hp
HOWARD DGA-15P
1939-41
DRAWN BY- PAUL R. MATT
from the files of
HISTORICAL AVIATION ALBUM
for
AMERICAN Aircraft MODELER
SCALE-ORIGINAL 1:24
DRAWING No. AM-52-A
PLATE 1 of 2

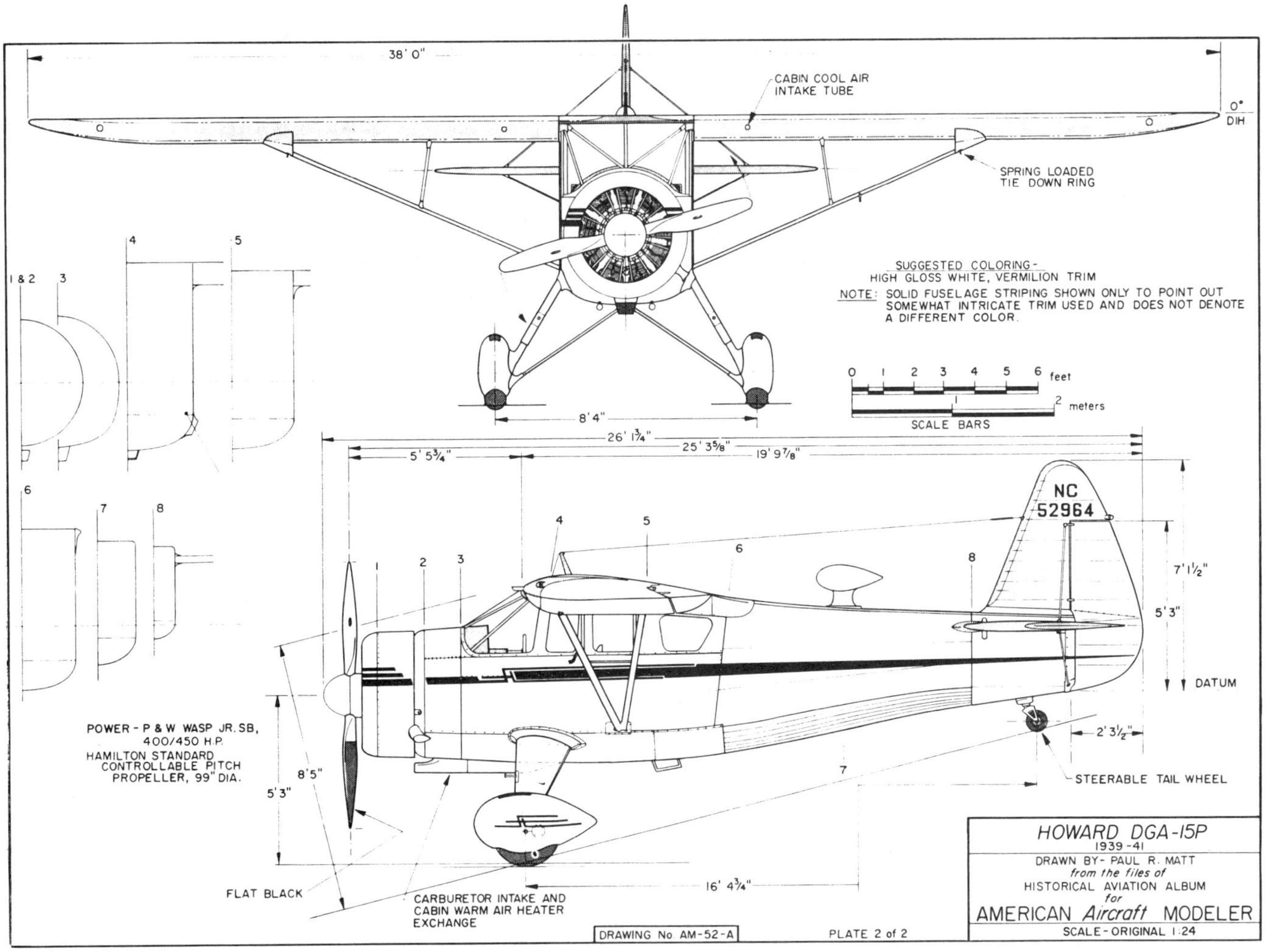
38' 0"
CABIN COOL AIR INTAKE TUBE
0° DIH
SPRING LOADED TIE DOWN RING
SUGGESTED COLORING-
HIGH GLOSS WHITE, VERMILION TRIM
NOTE: SOLID FUSELAGE STRIPING SHOWN ONLY TO POINT OUT SOMEWHAT INTRICATE TRIM USED AND DOES NOT DENOTE A DIFFERENT COLOR.
0 1 2 3 4 5 6 feet
1 2 meters
SCALE BARS
8'4"
1 & 2
3
4
5
6
7
8
26' 1¾"
25' 3⅝"
5' 5¾"
19' 9⅞"
NC
52964
7'1½"
5'3"
DATUM
2' 3½"
STEERABLE TAIL WHEEL
POWER - P & W WASP JR. SB, 400/450 H.P.
HAMILTON STANDARD CONTROLLABLE PITCH PROPELLER, 99" DIA.
8'5"
5'3"
FLAT BLACK
CARBURETOR INTAKE AND CABIN WARM AIR HEATER EXCHANGE
16' 4¾"
HOWARD DGA-15P
1939-41
DRAWN BY- PAUL R. MATT
from the files of
HISTORICAL AVIATION ALBUM
for
AMERICAN Aircraft MODELER
SCALE-ORIGINAL 1:24
DRAWING No. AM-52-A
PLATE 2 of 2

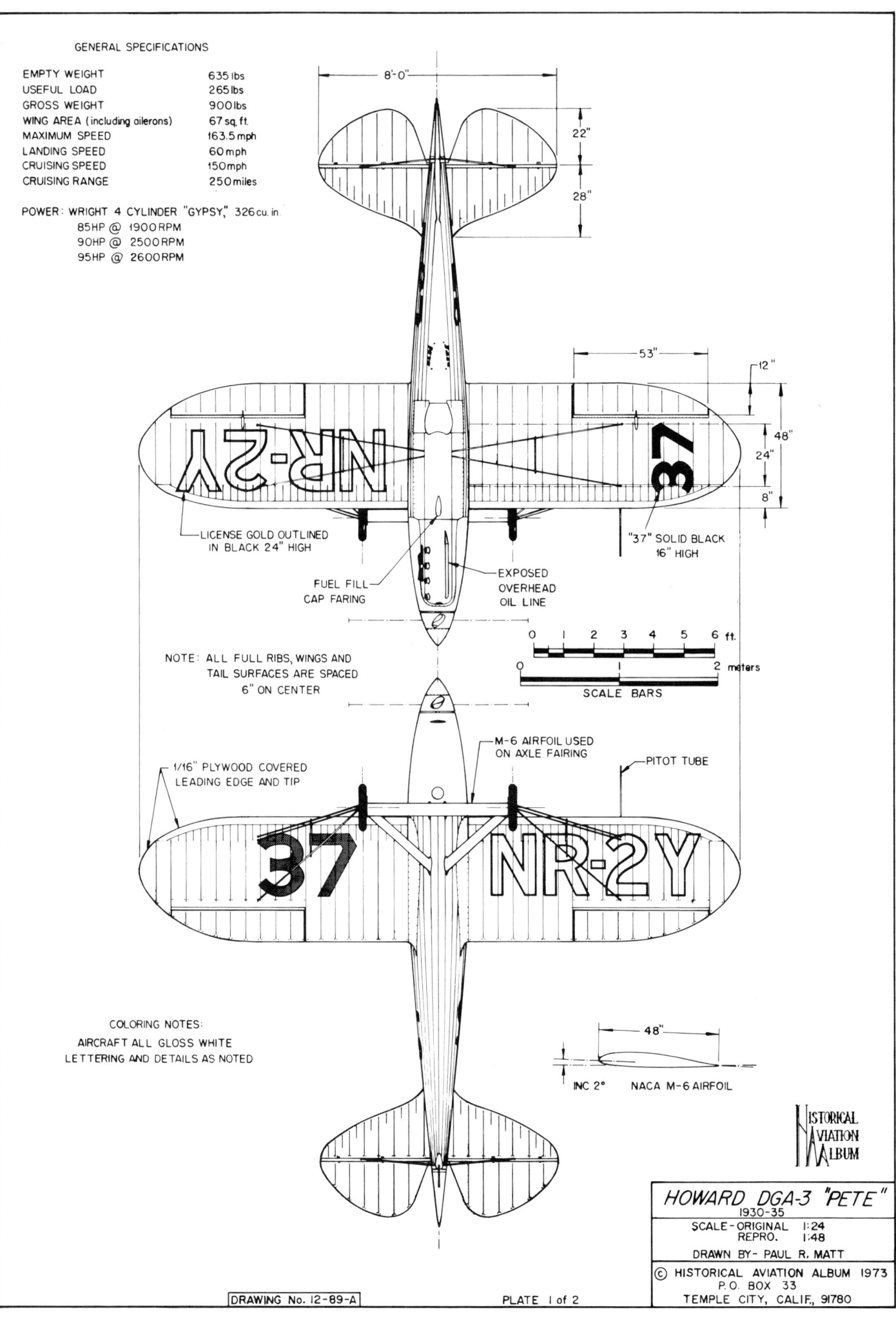

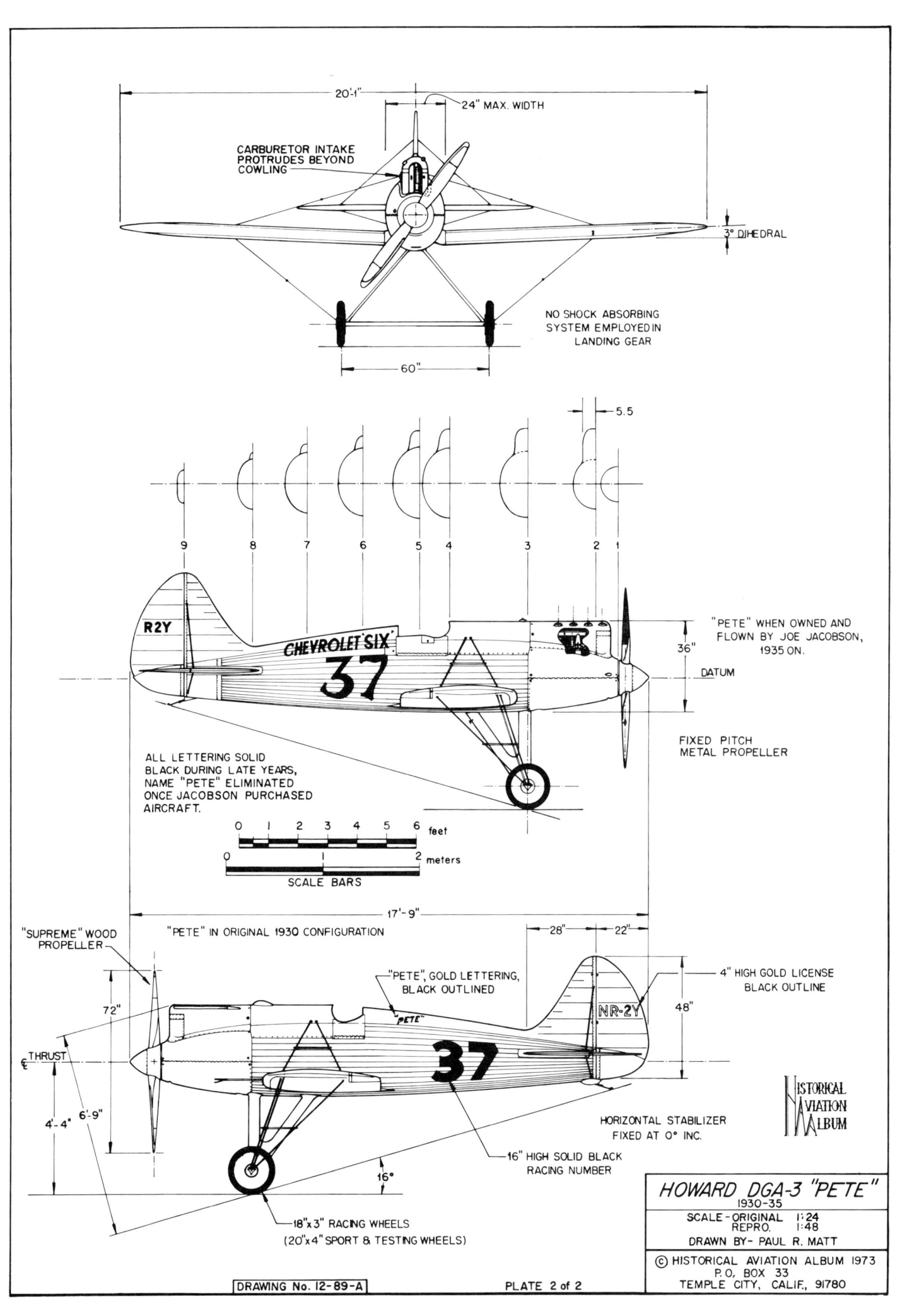

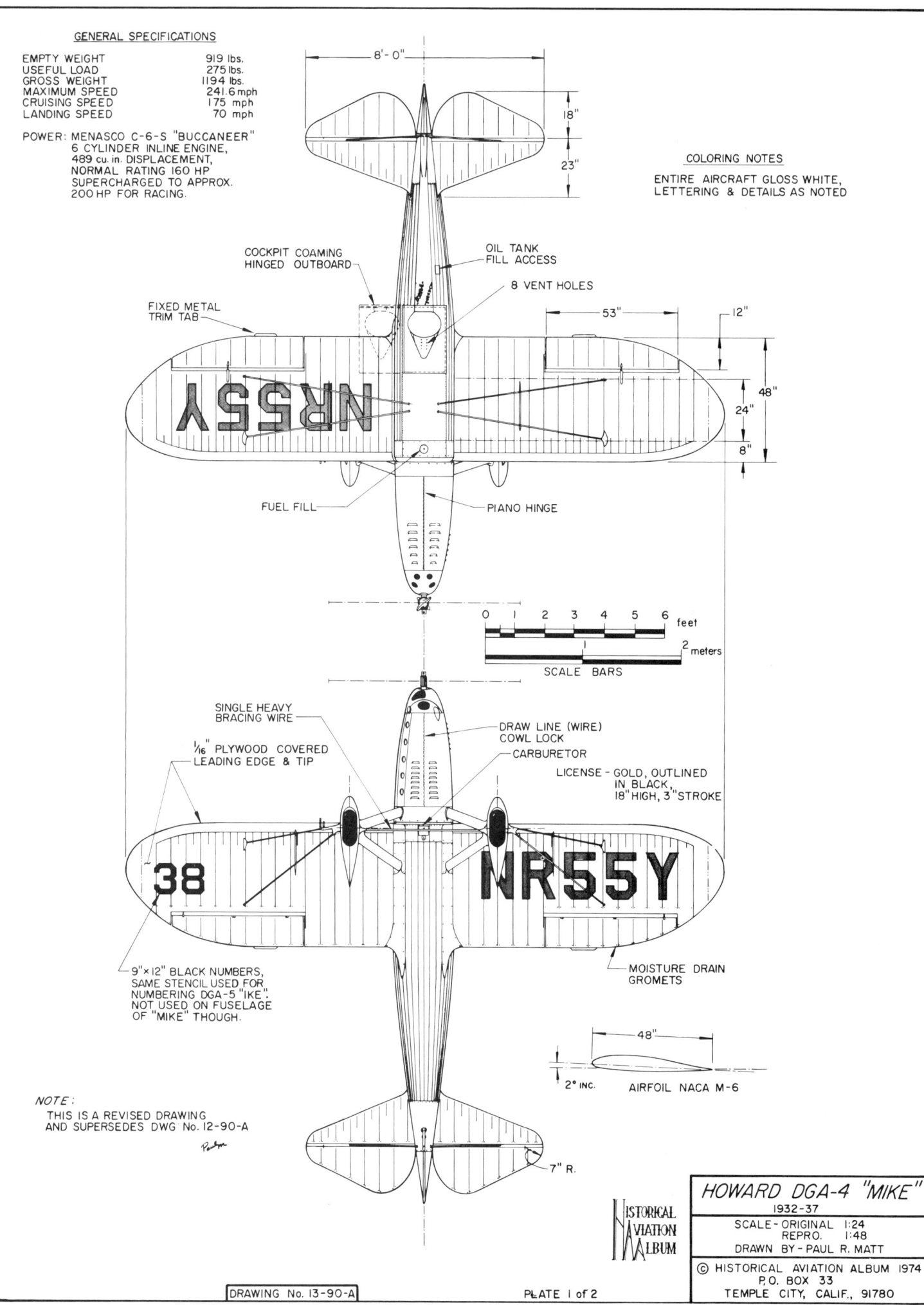

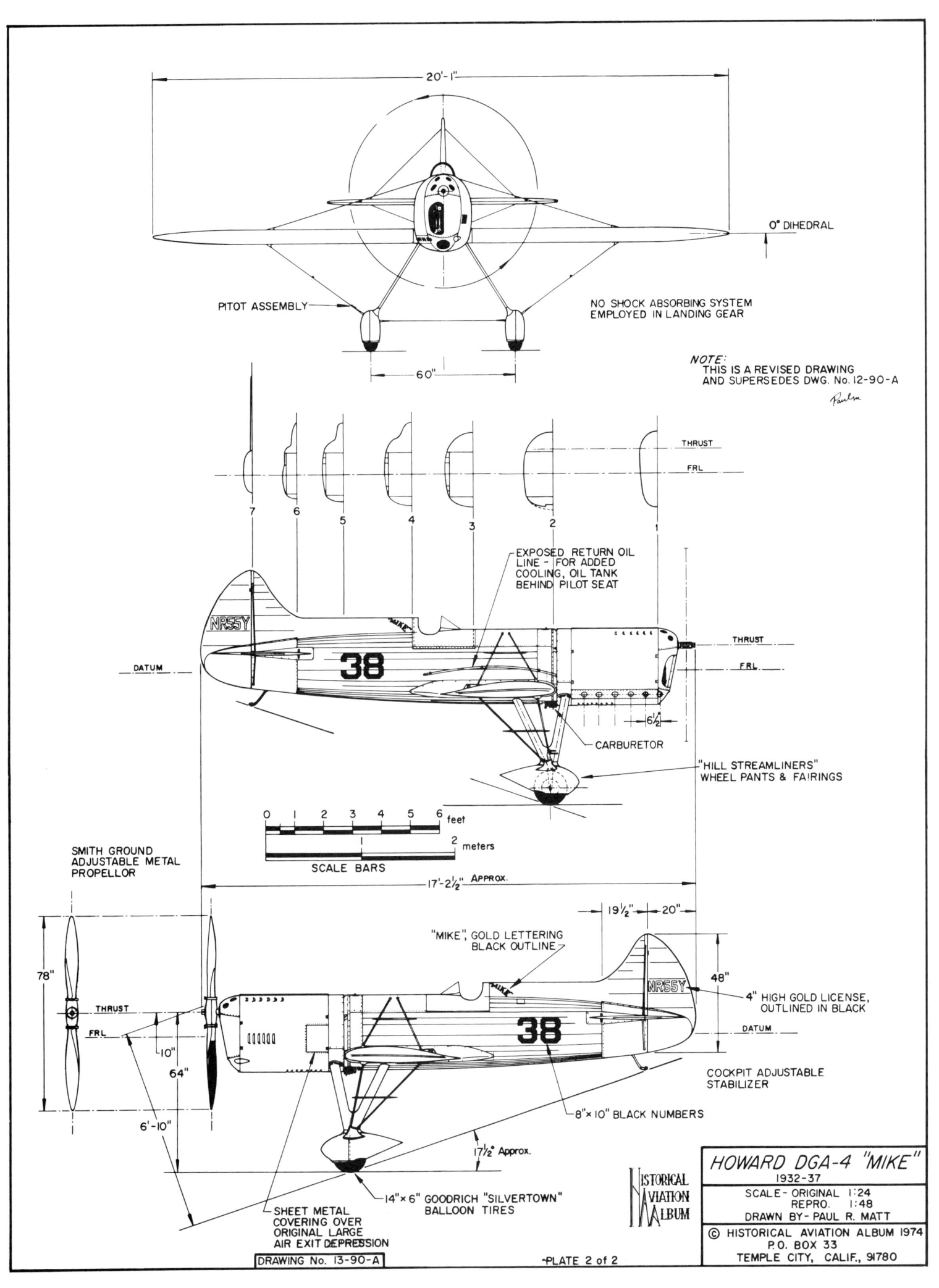

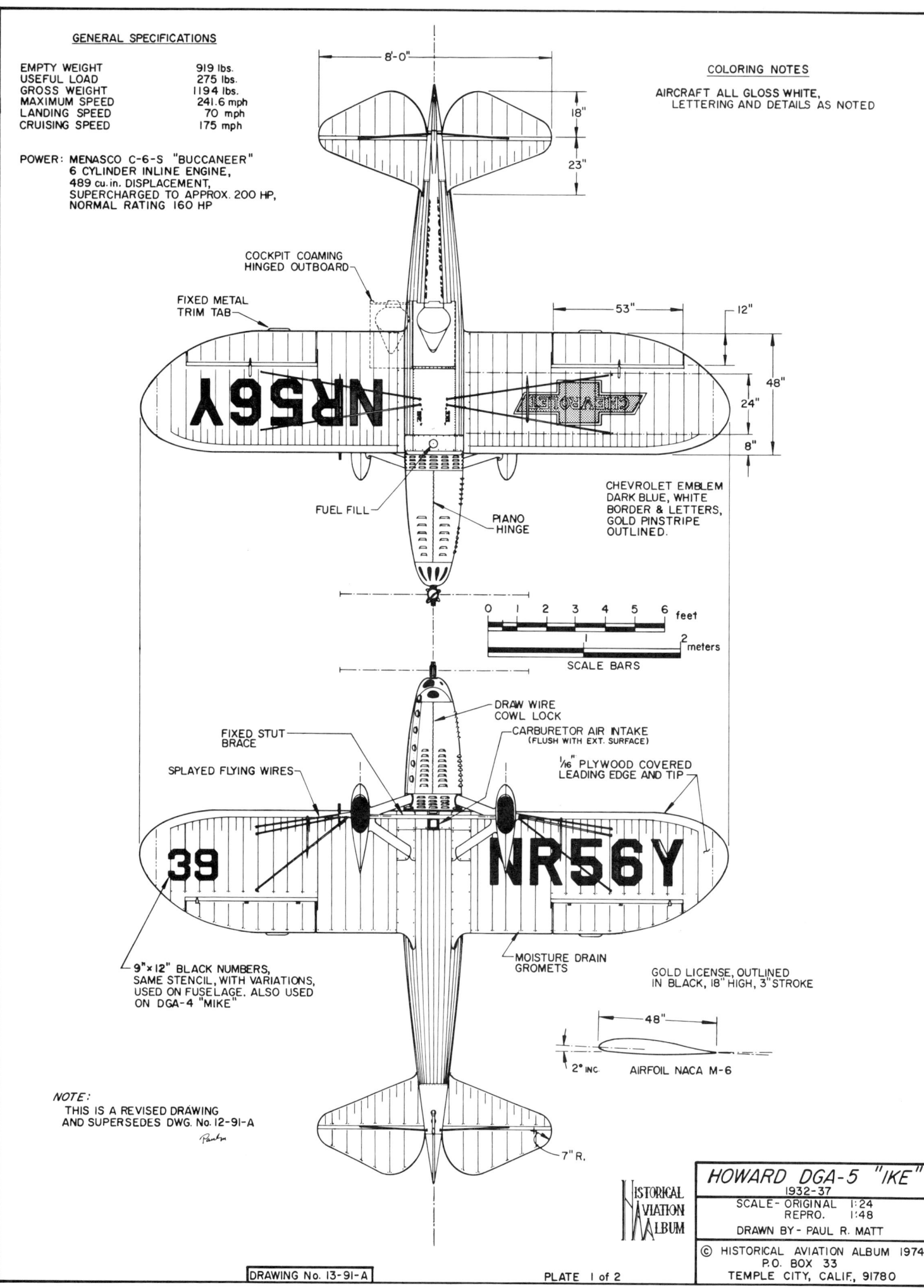

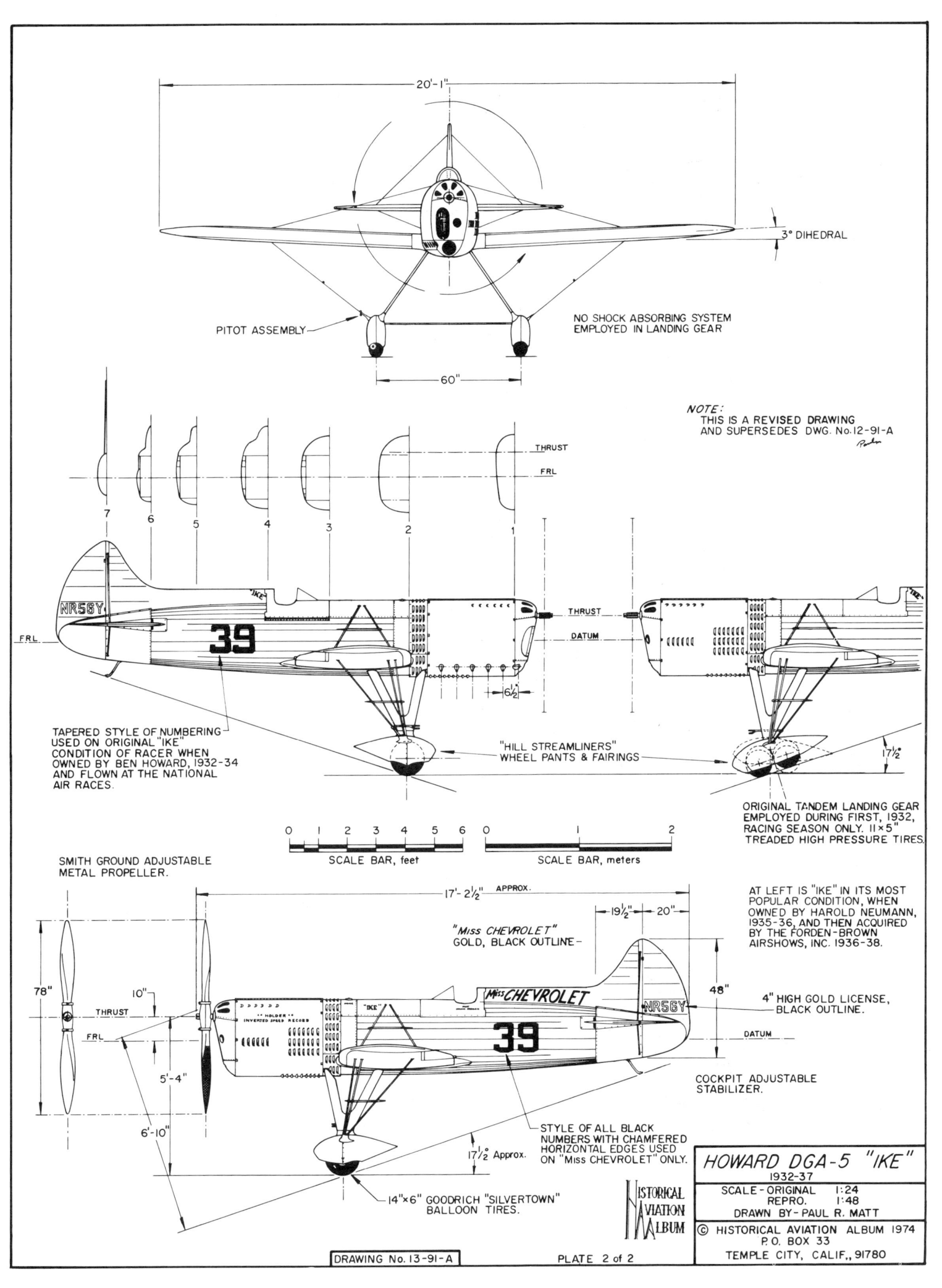

Howard DGA-6 Mister Mulligan

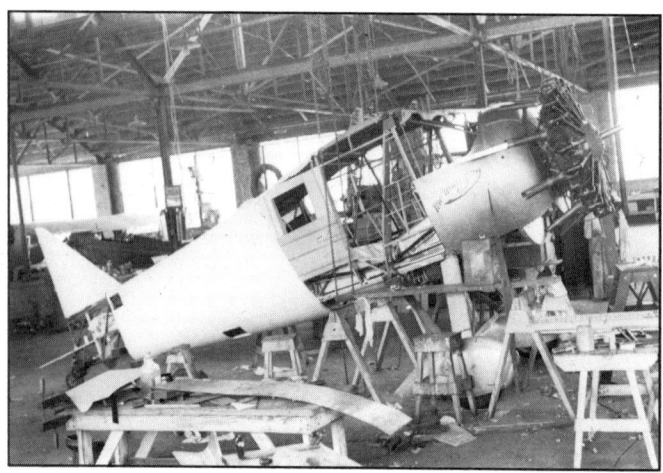

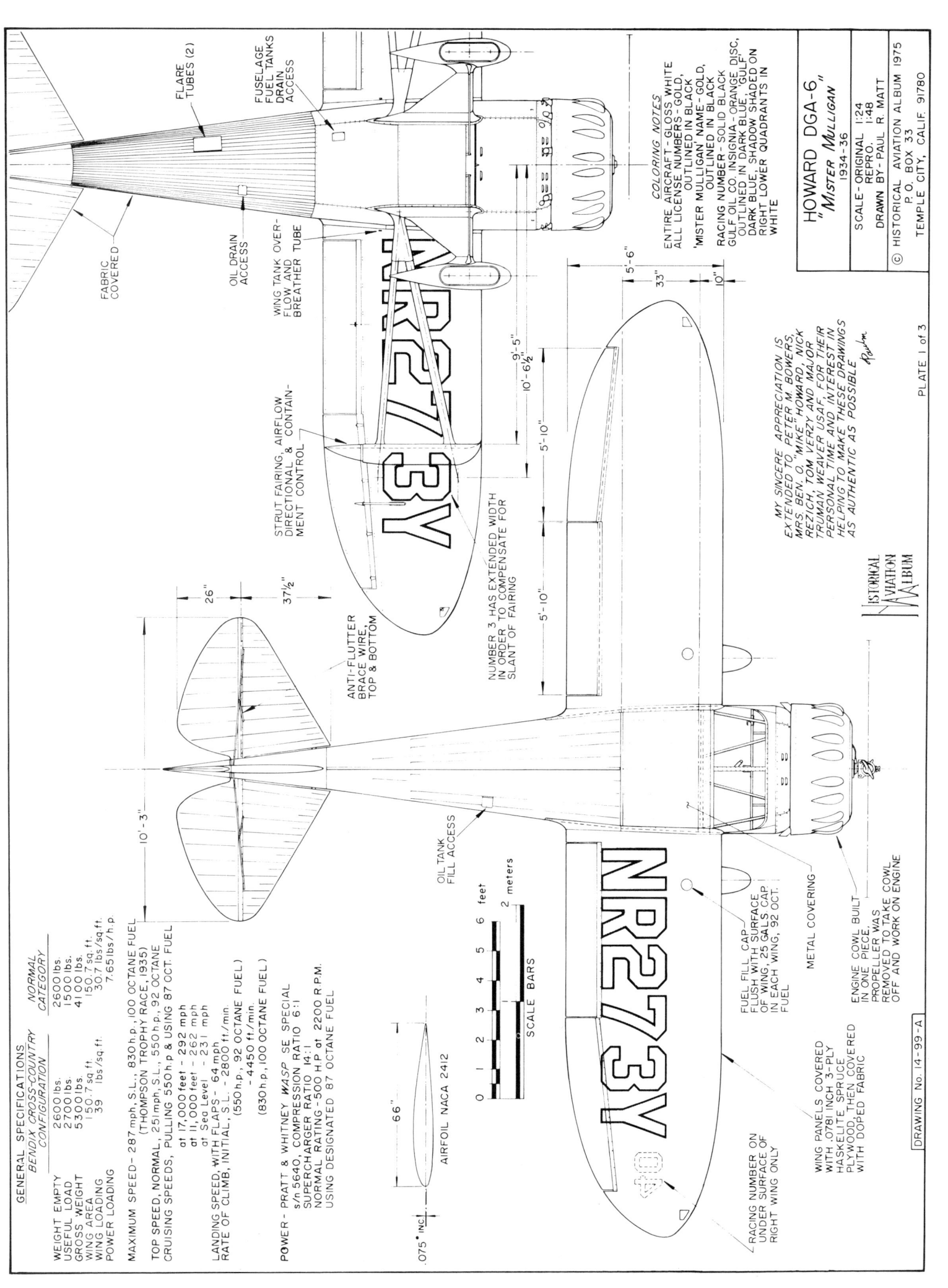

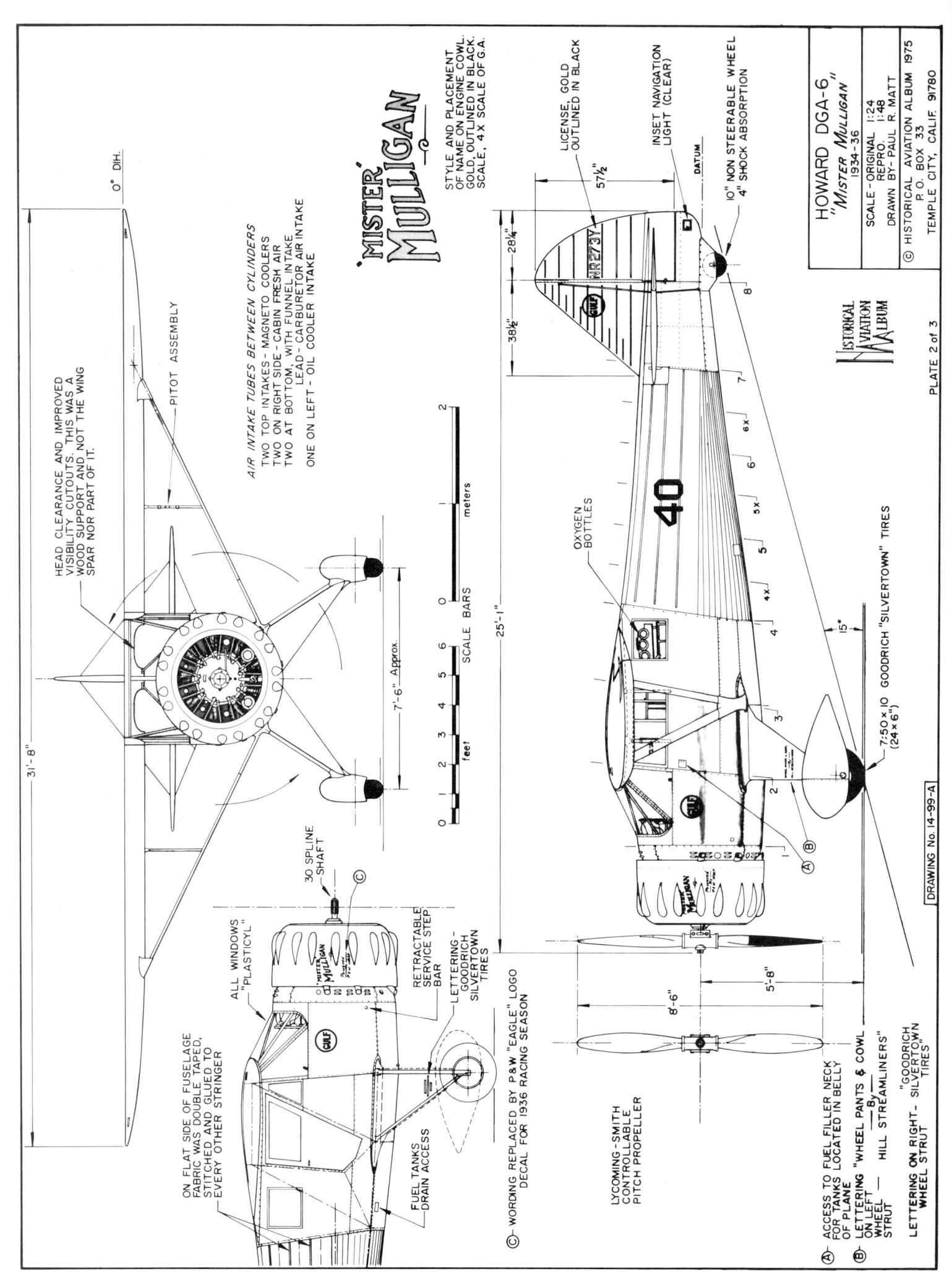

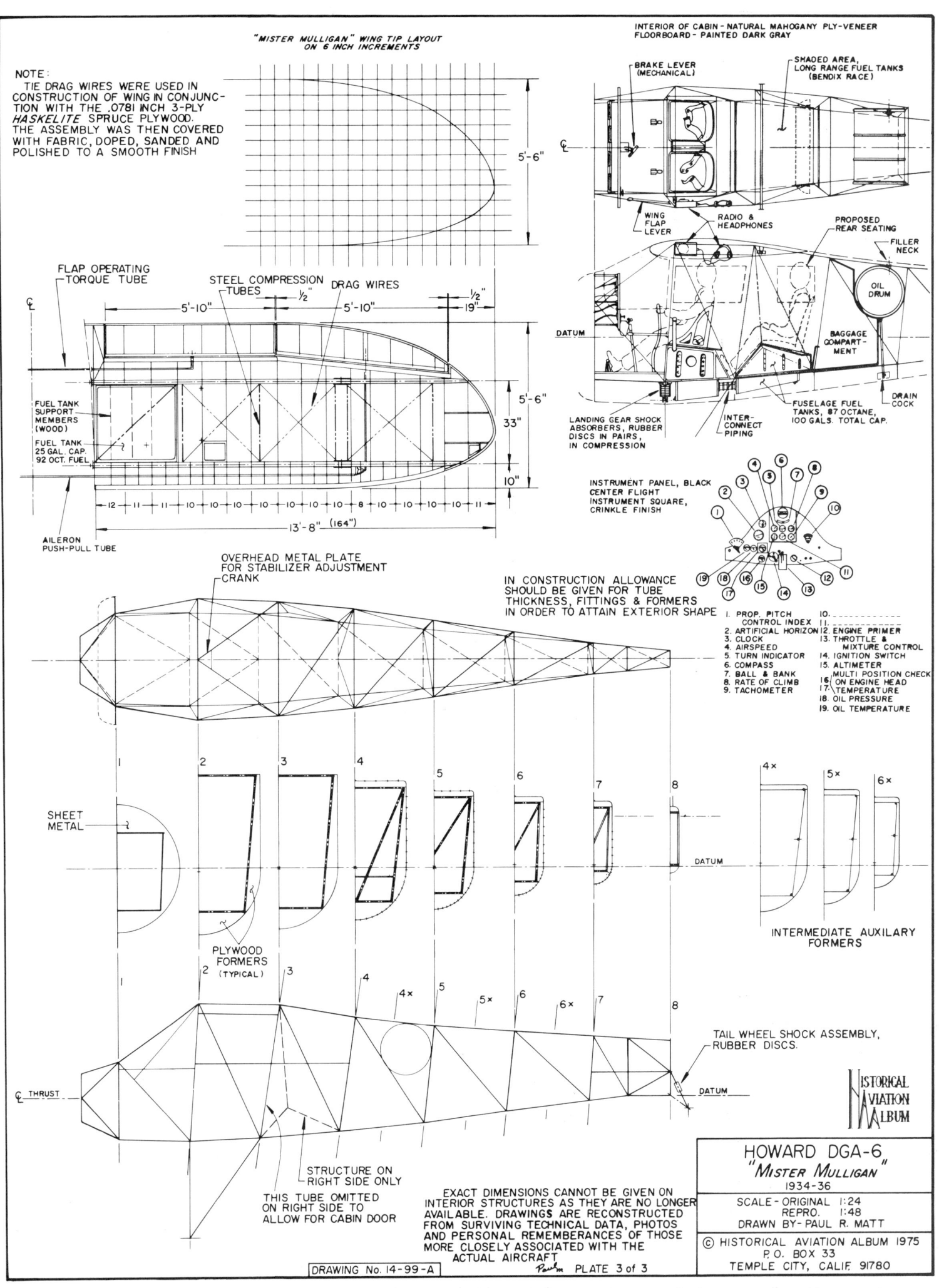

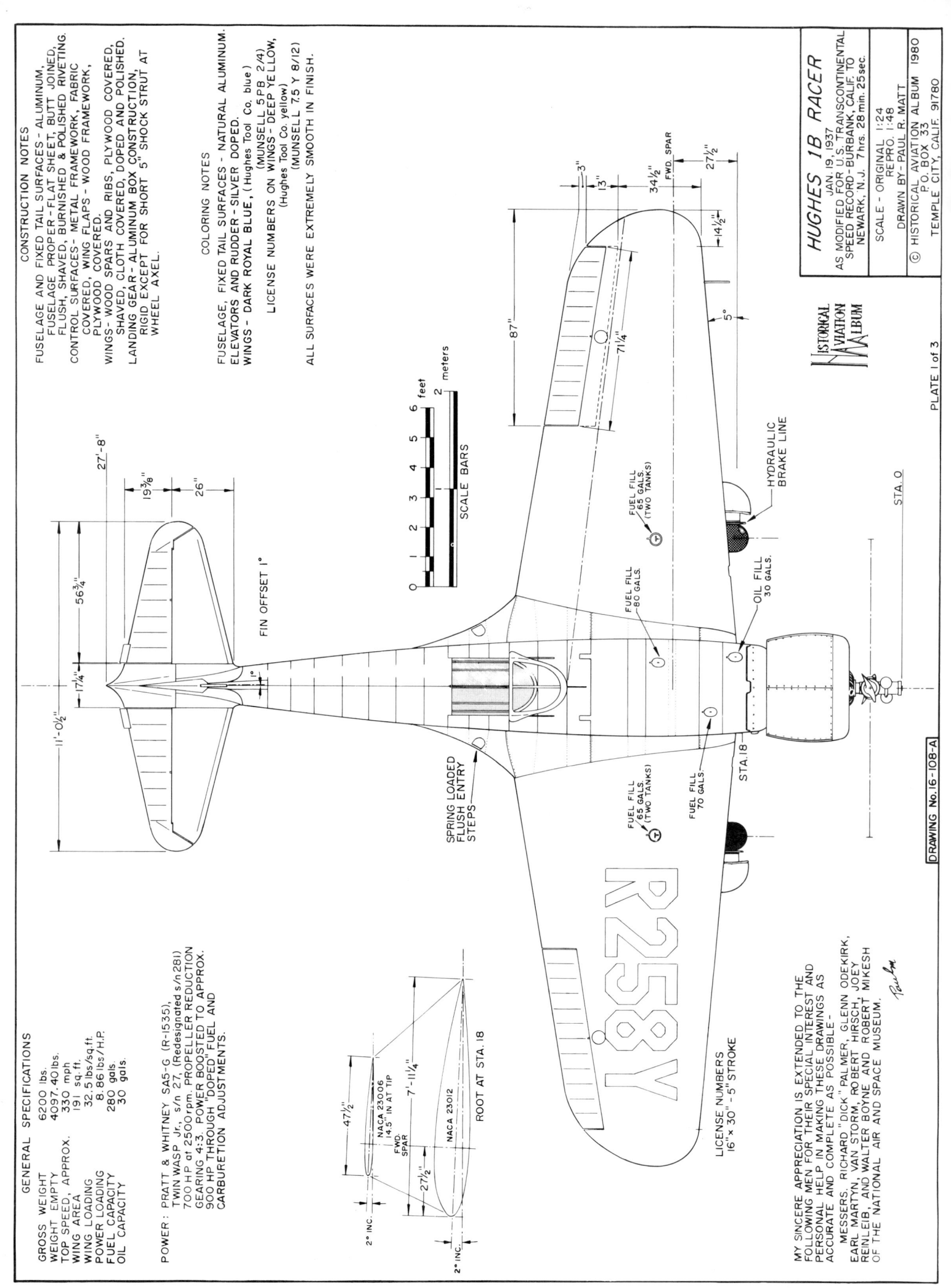

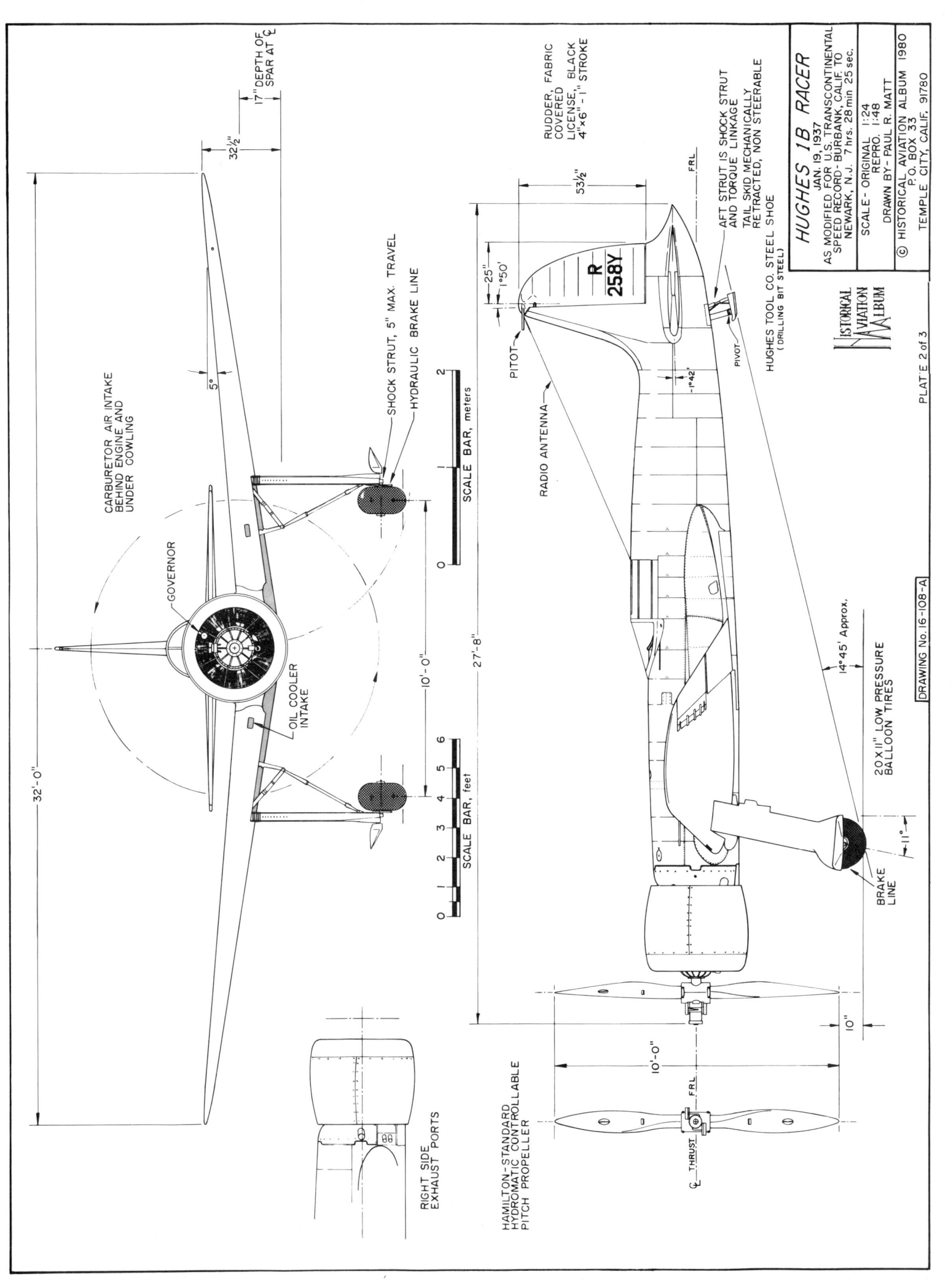

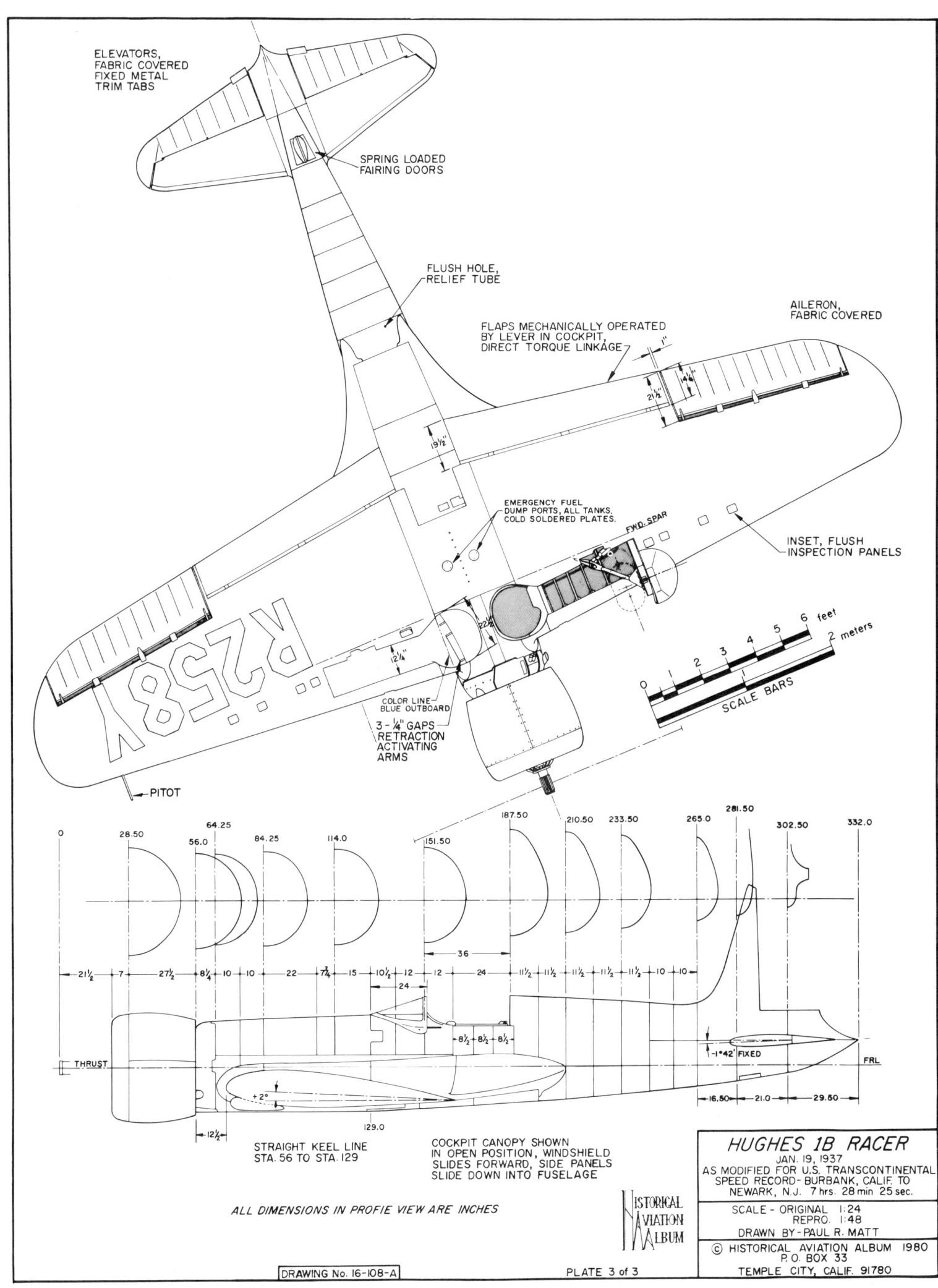

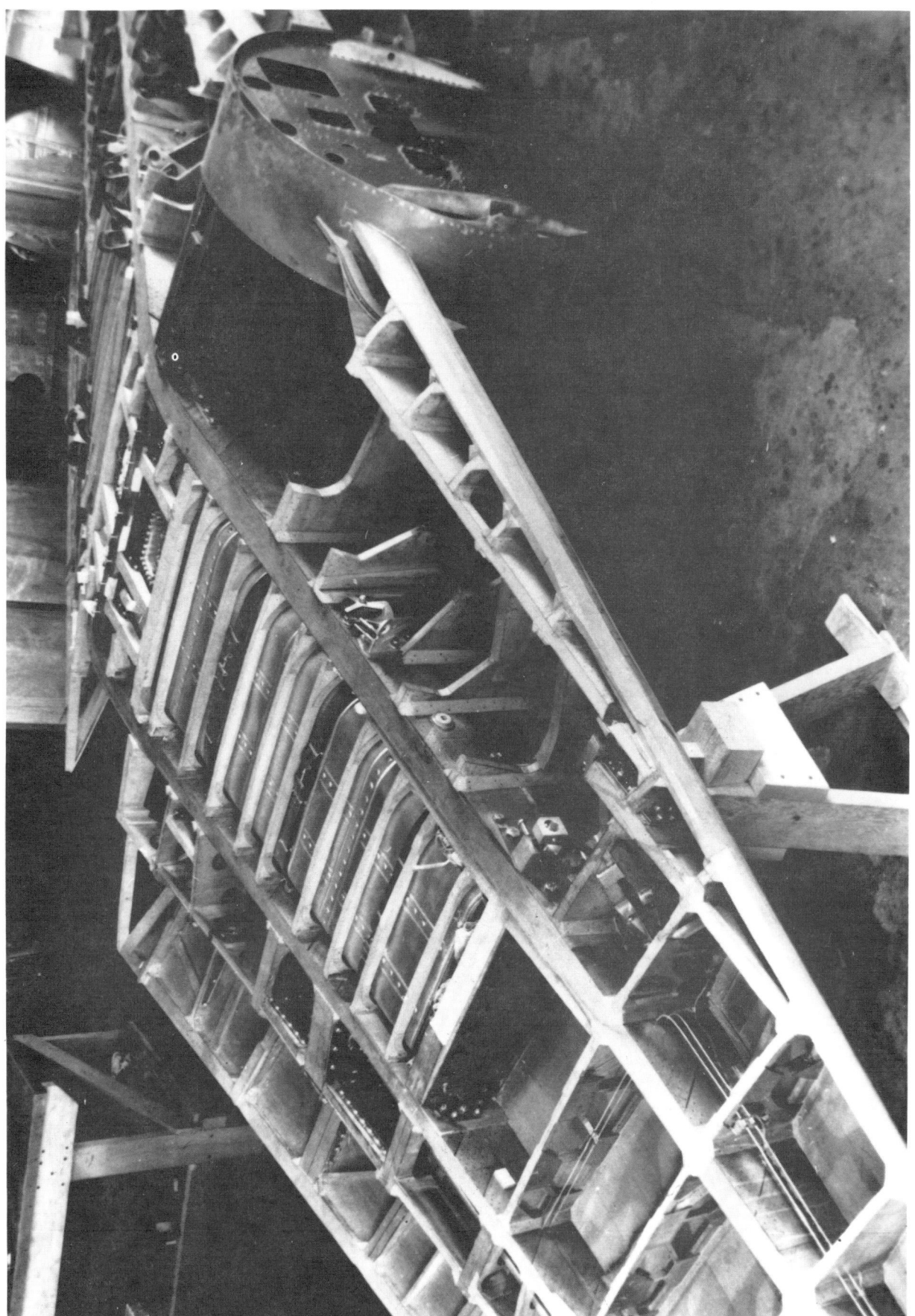

Hughes 1B Racer, "Longwing" under construction in 1937.

Hughes Shortwing Racer

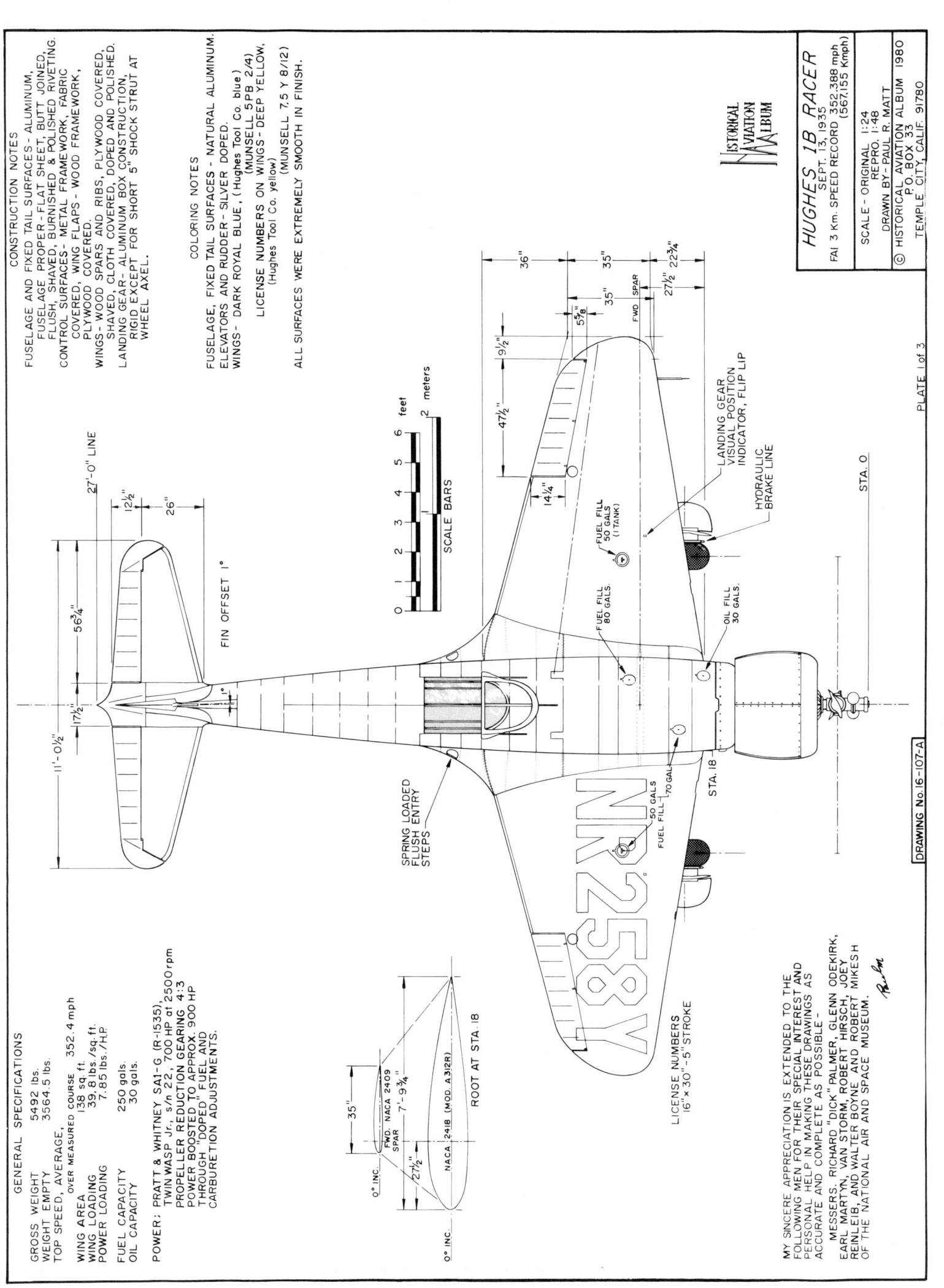

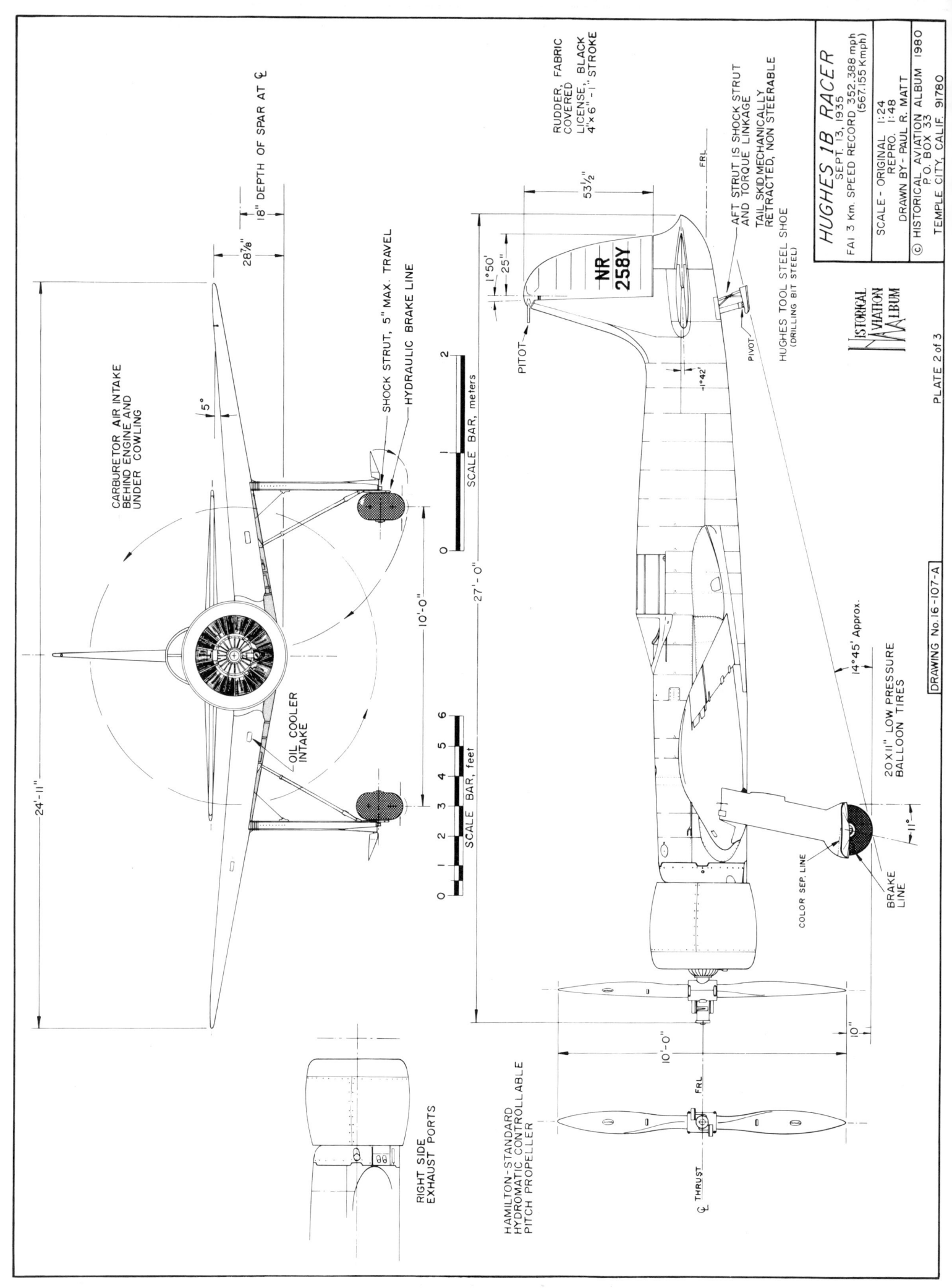

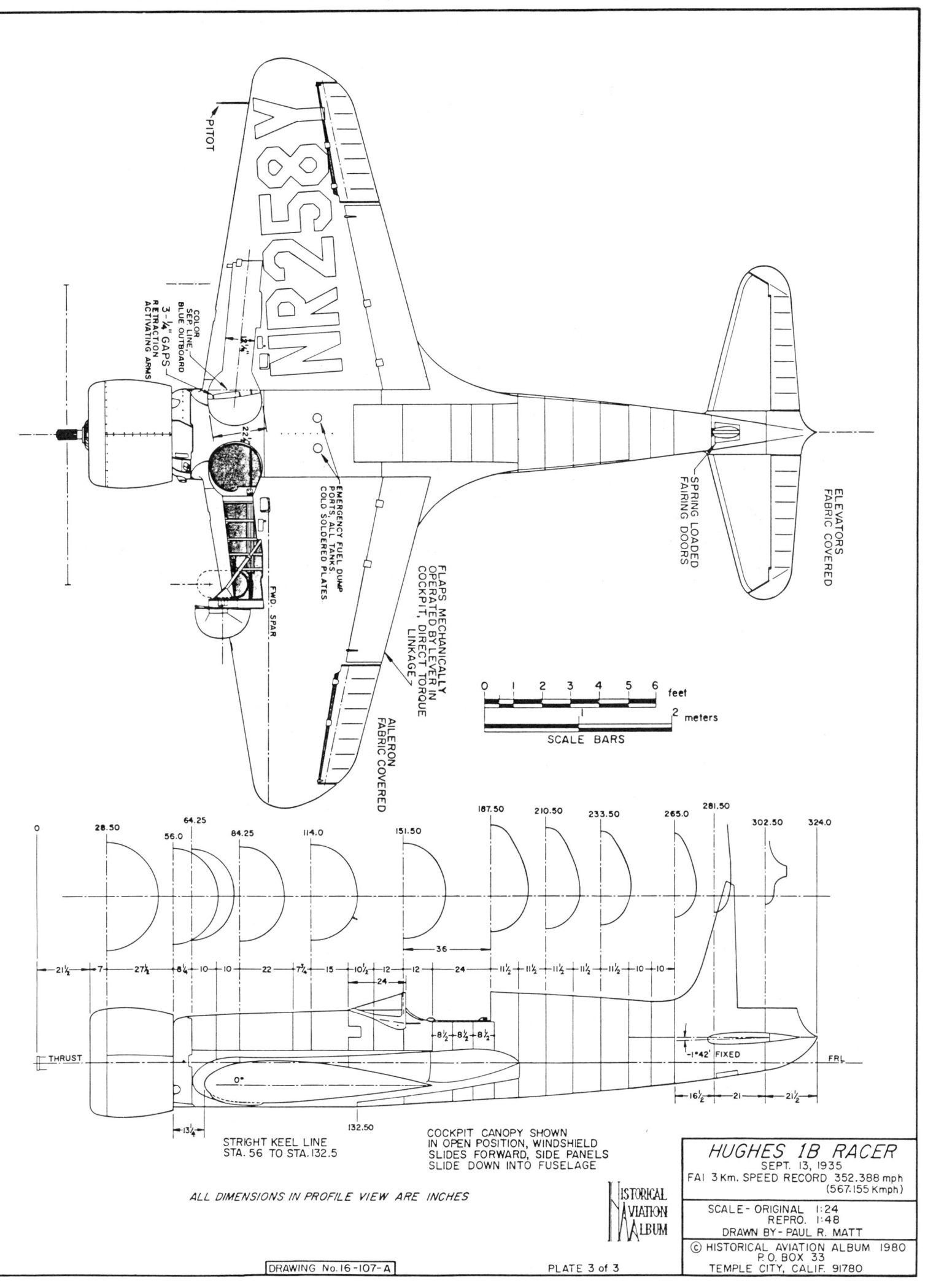

Laird LC-DW-300 "Solution"

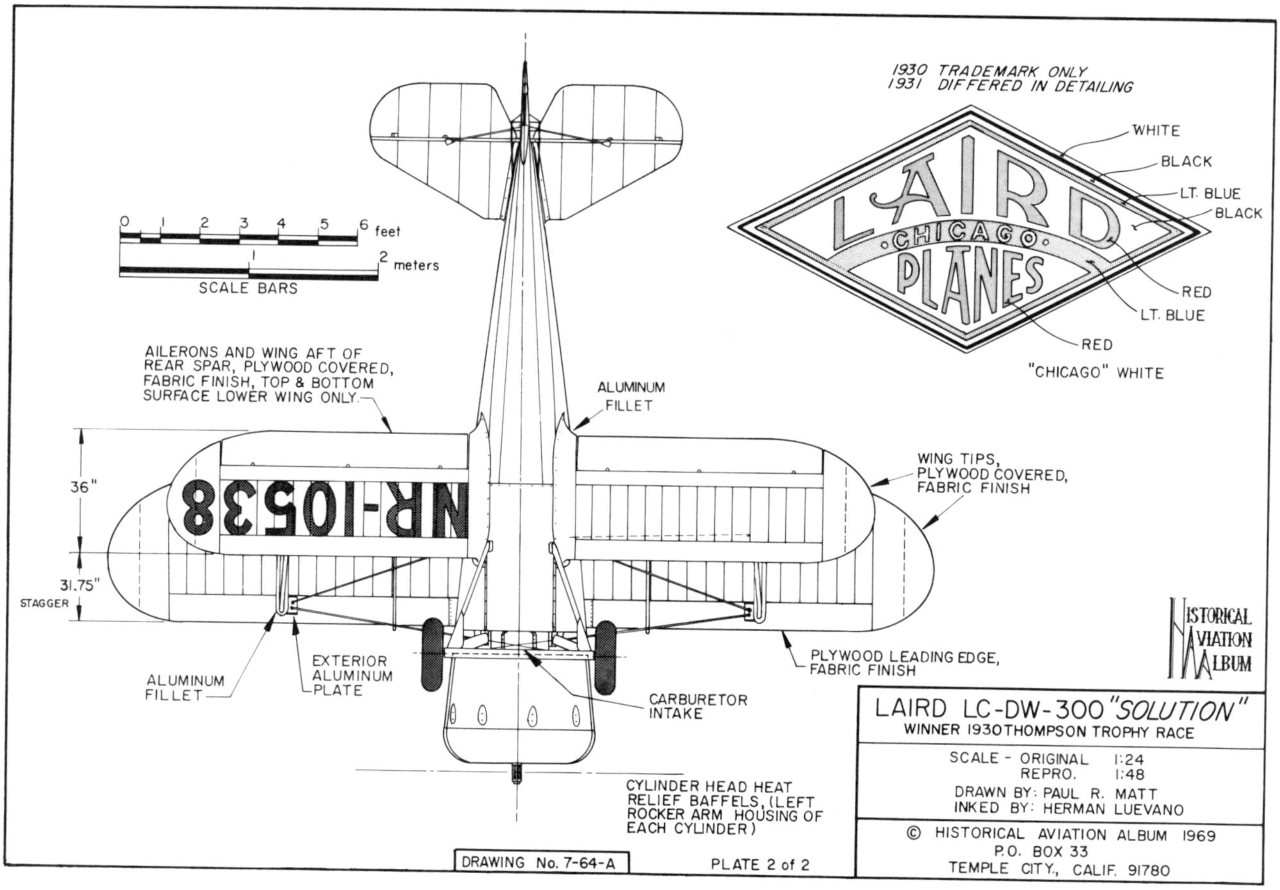

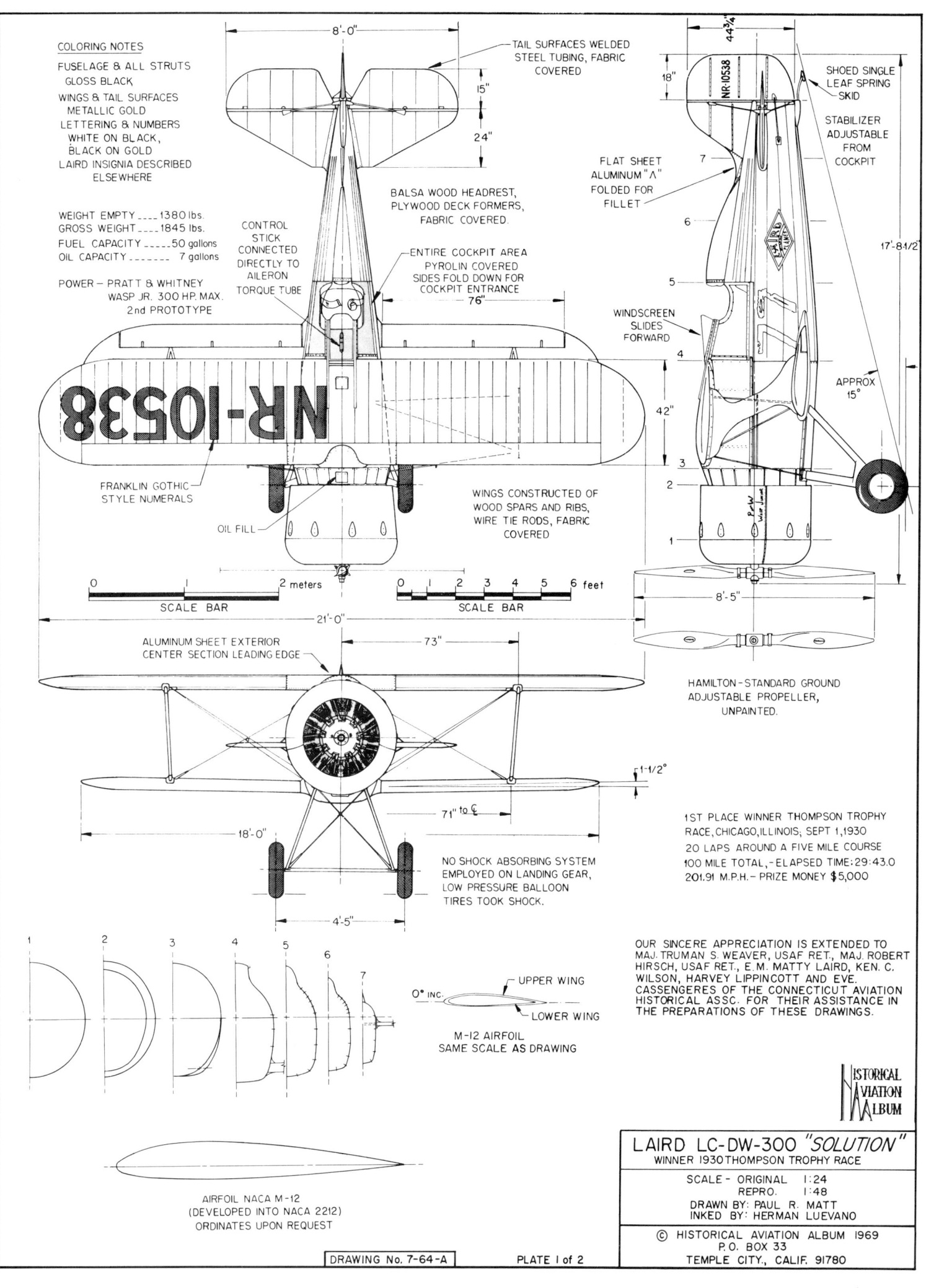

Laird LC-DW-500 "Super Solution"

Jimmy Doolittle in front of Laird Super Solution, September, 1931.

Laird Super Solution

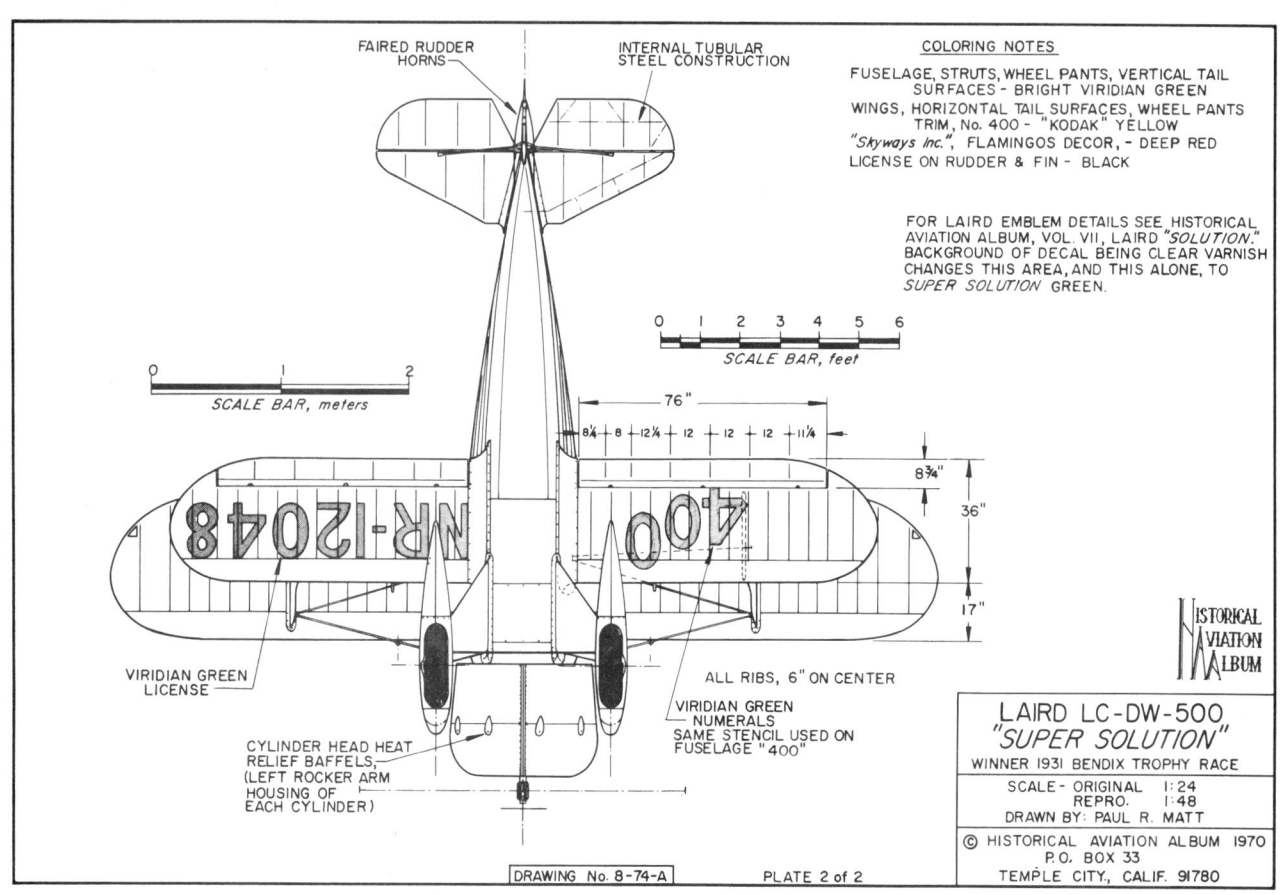

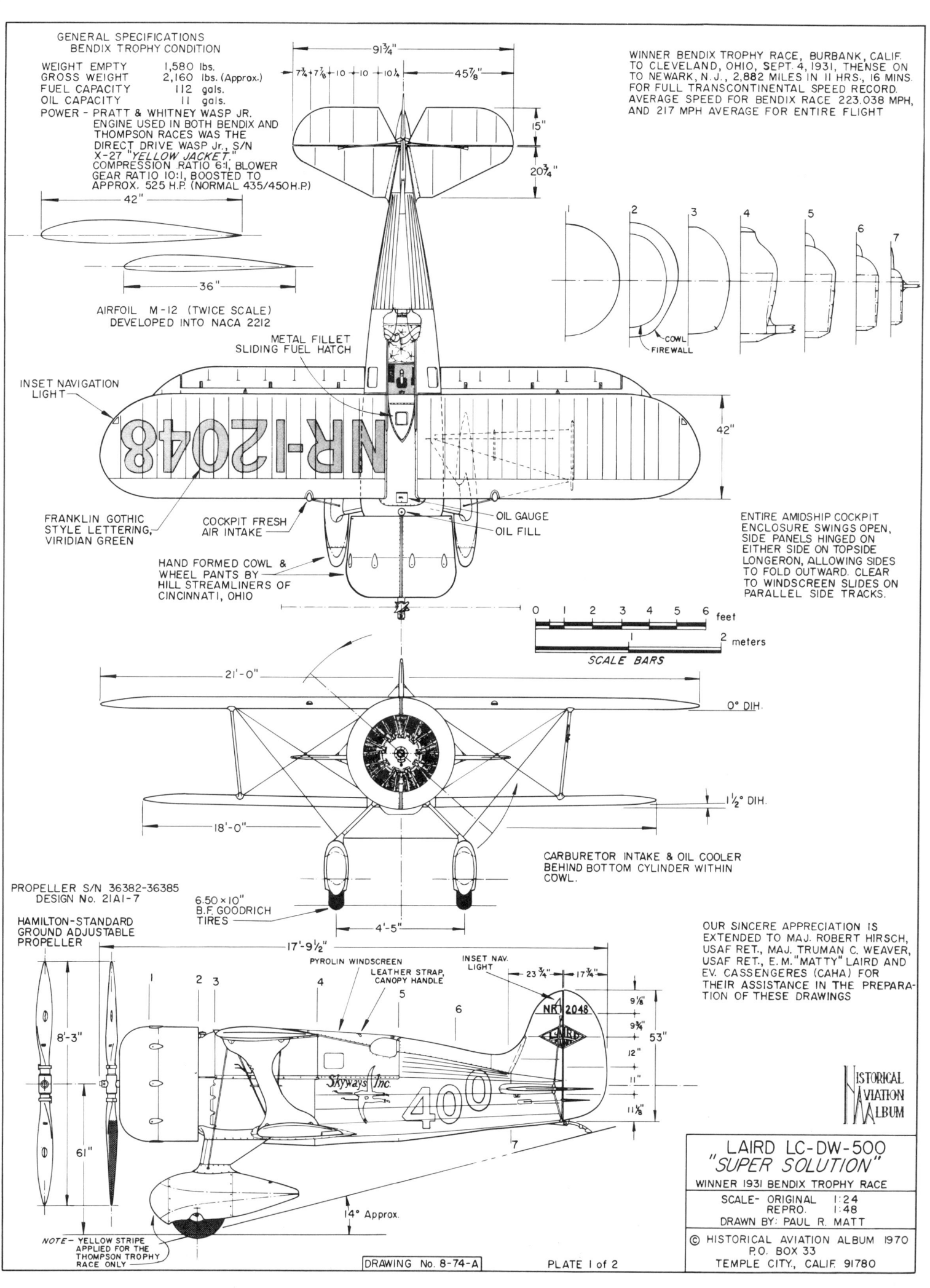

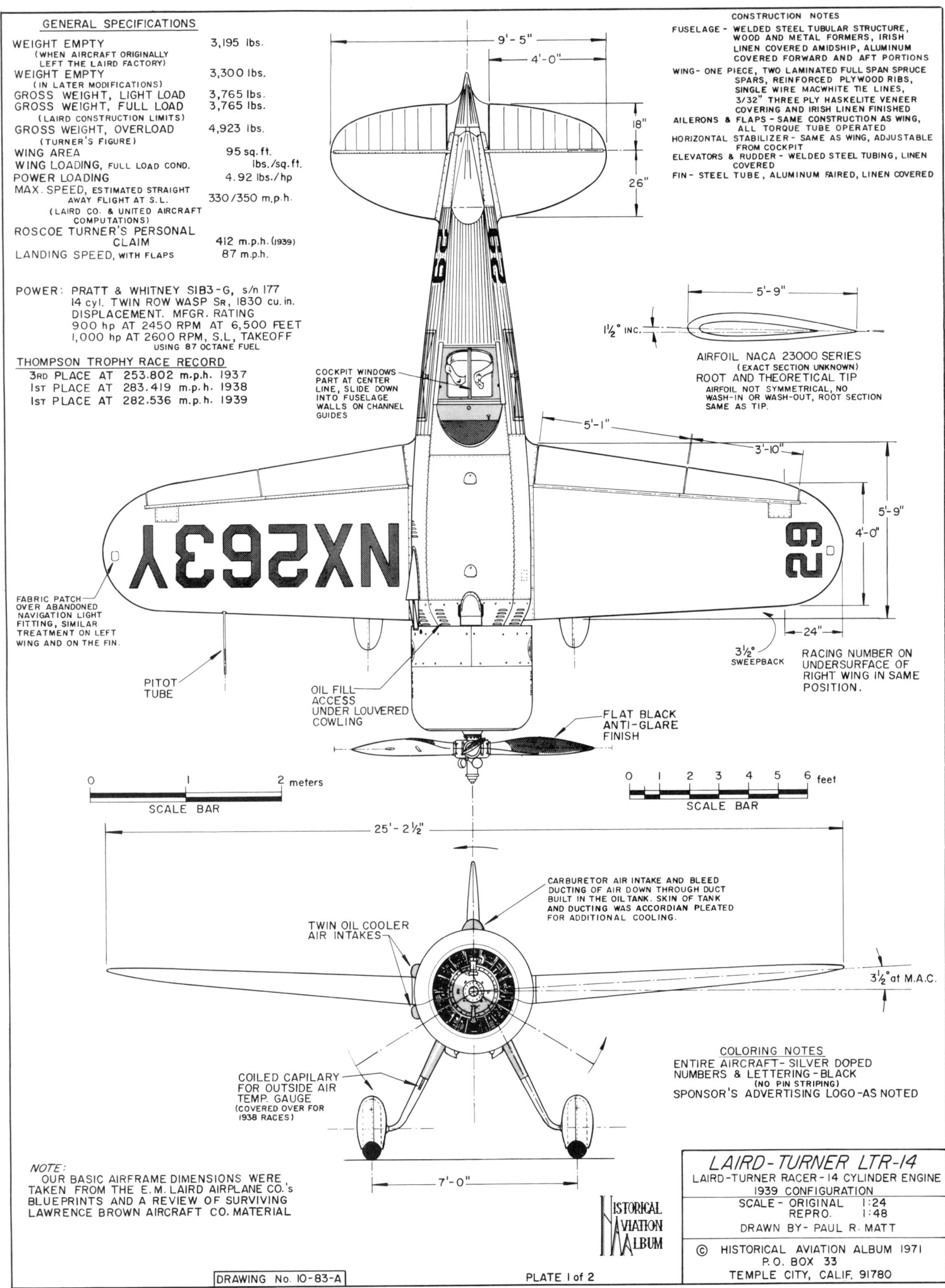

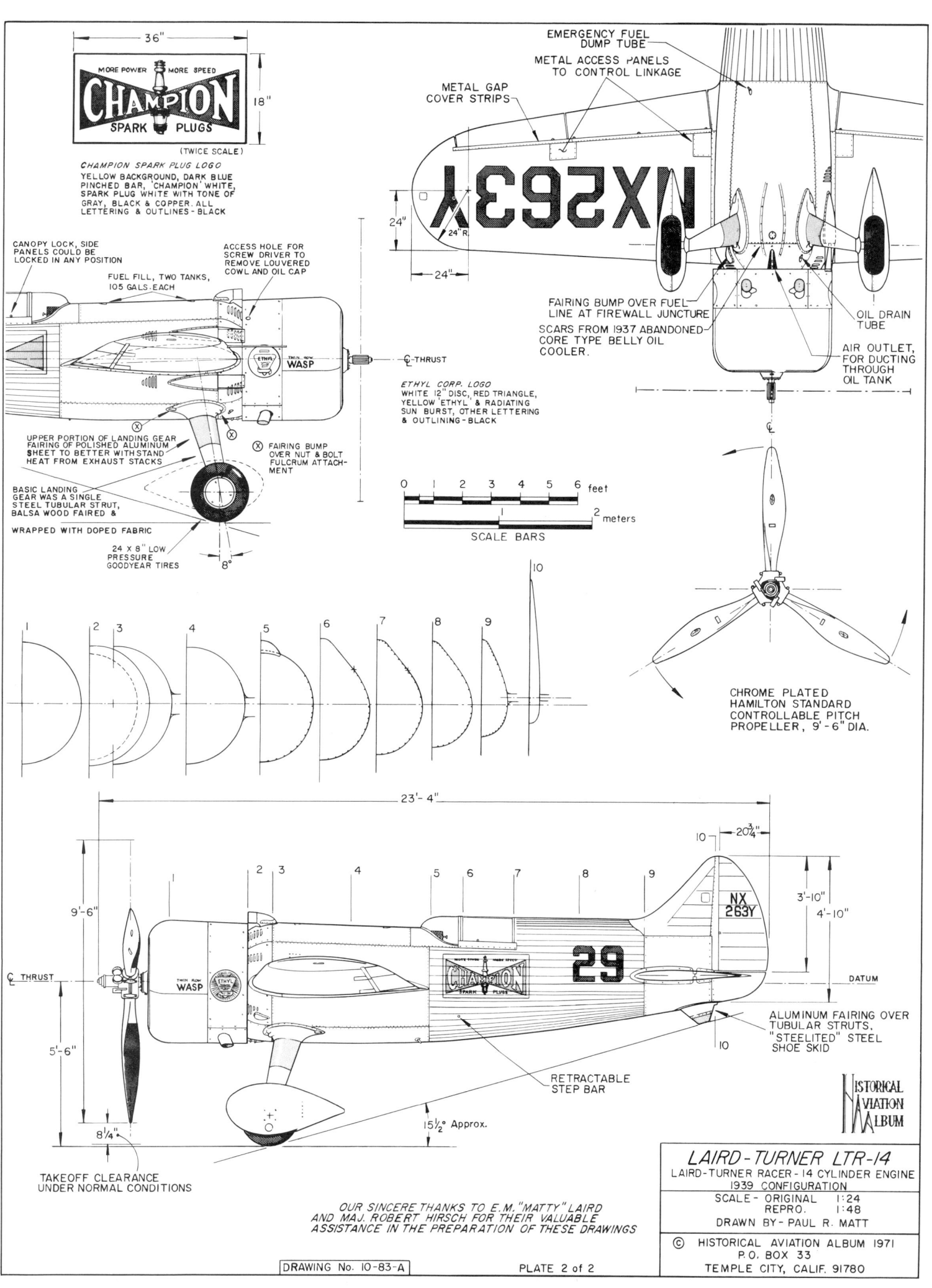

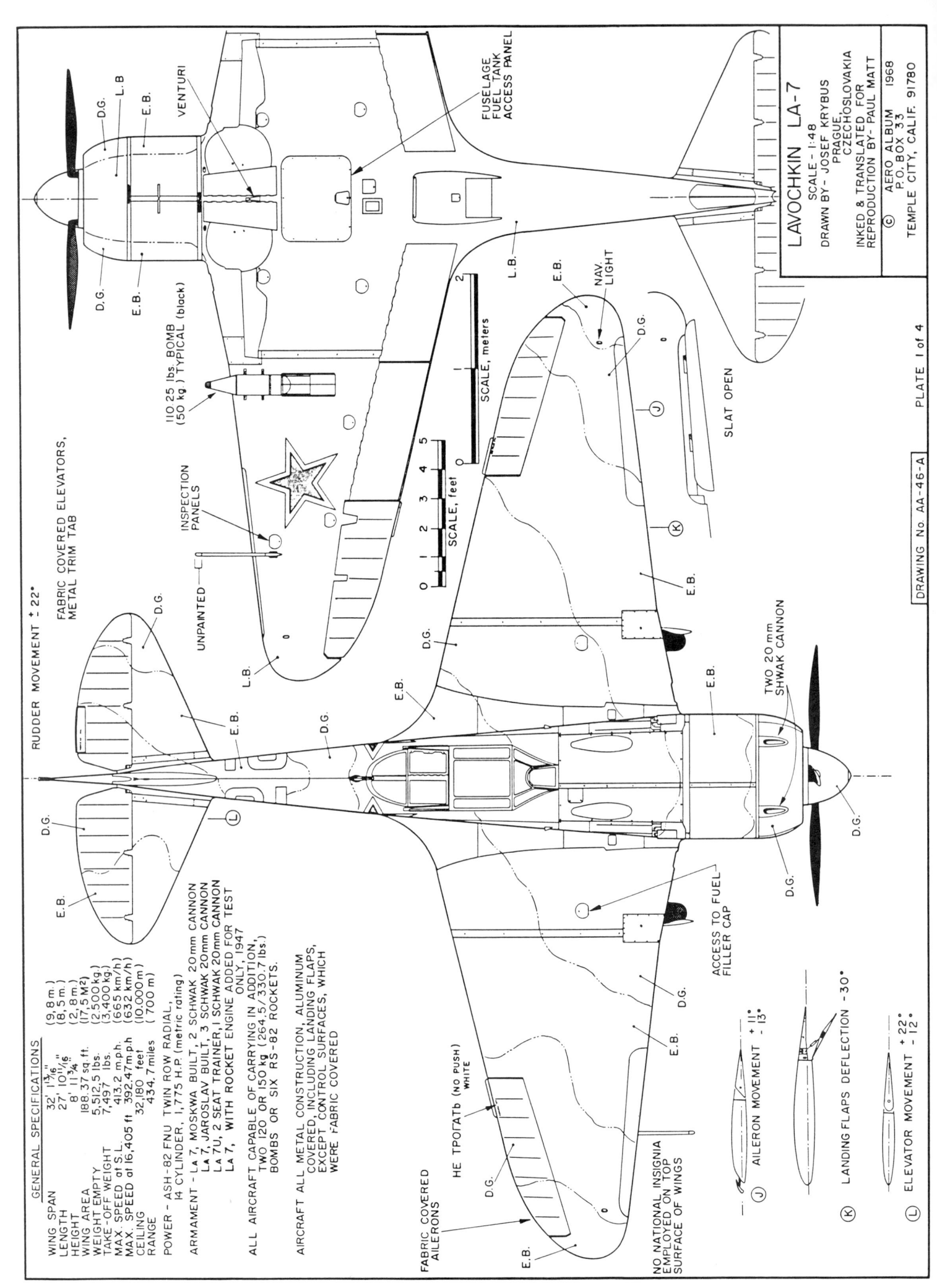

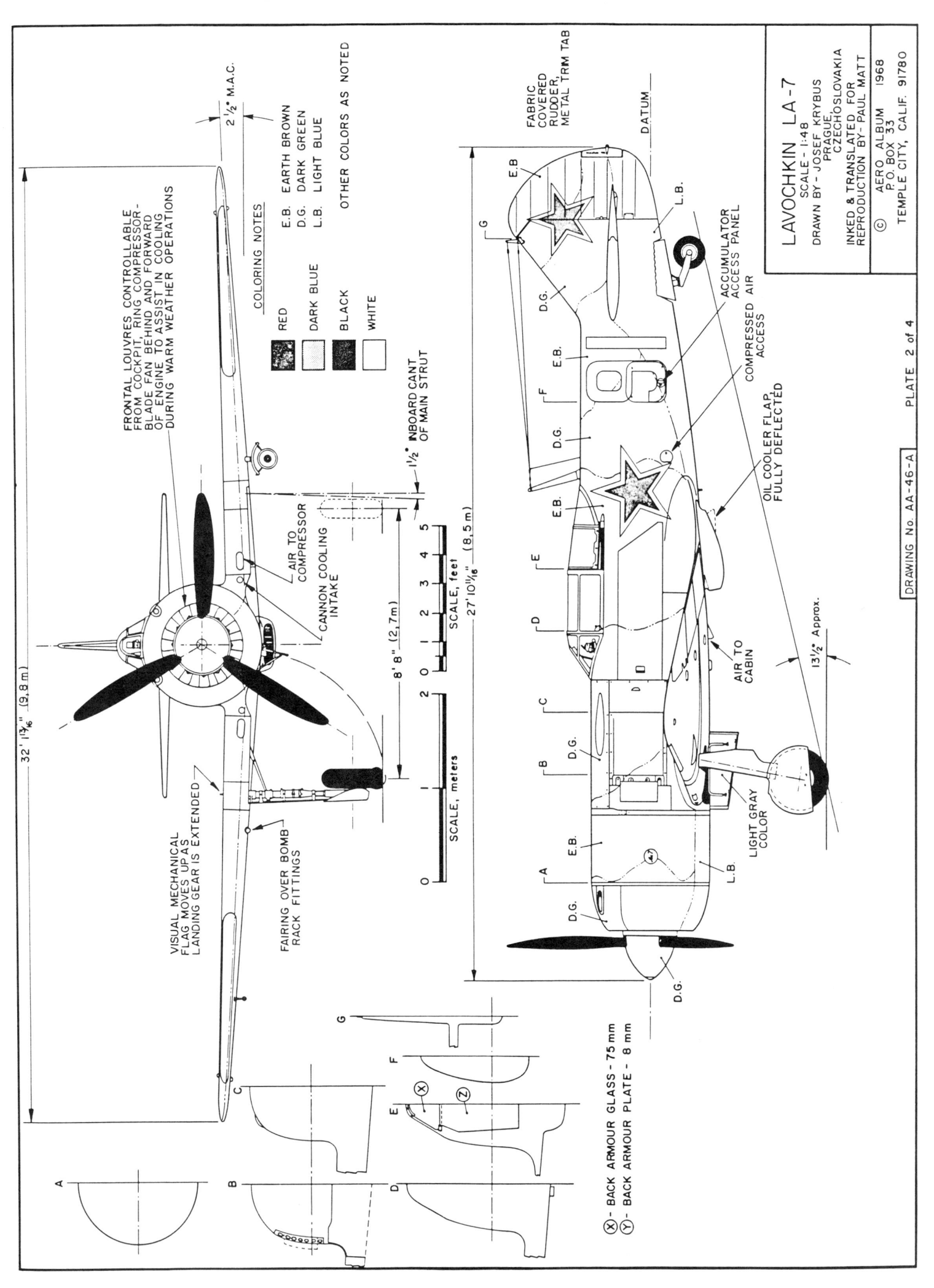

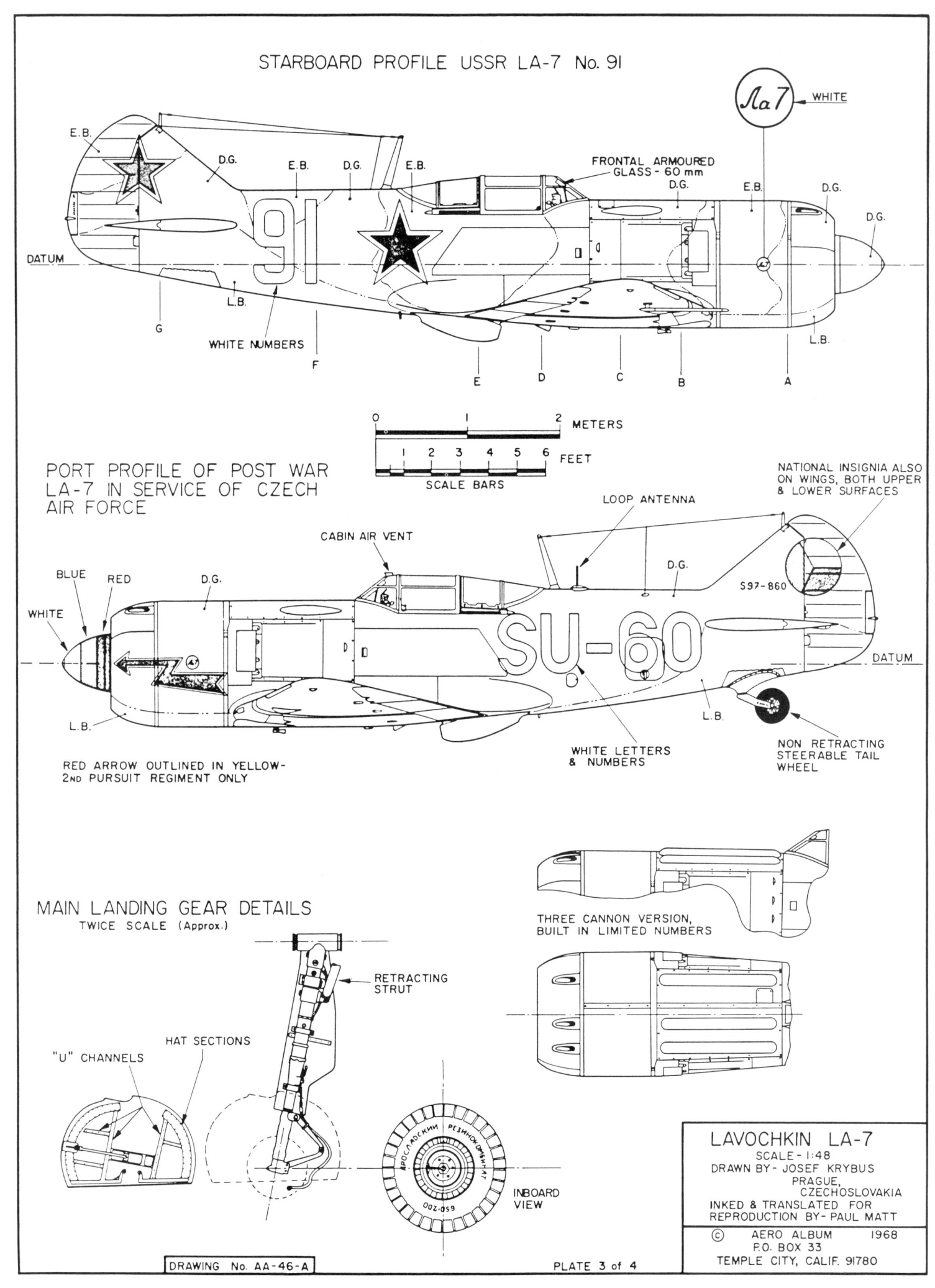

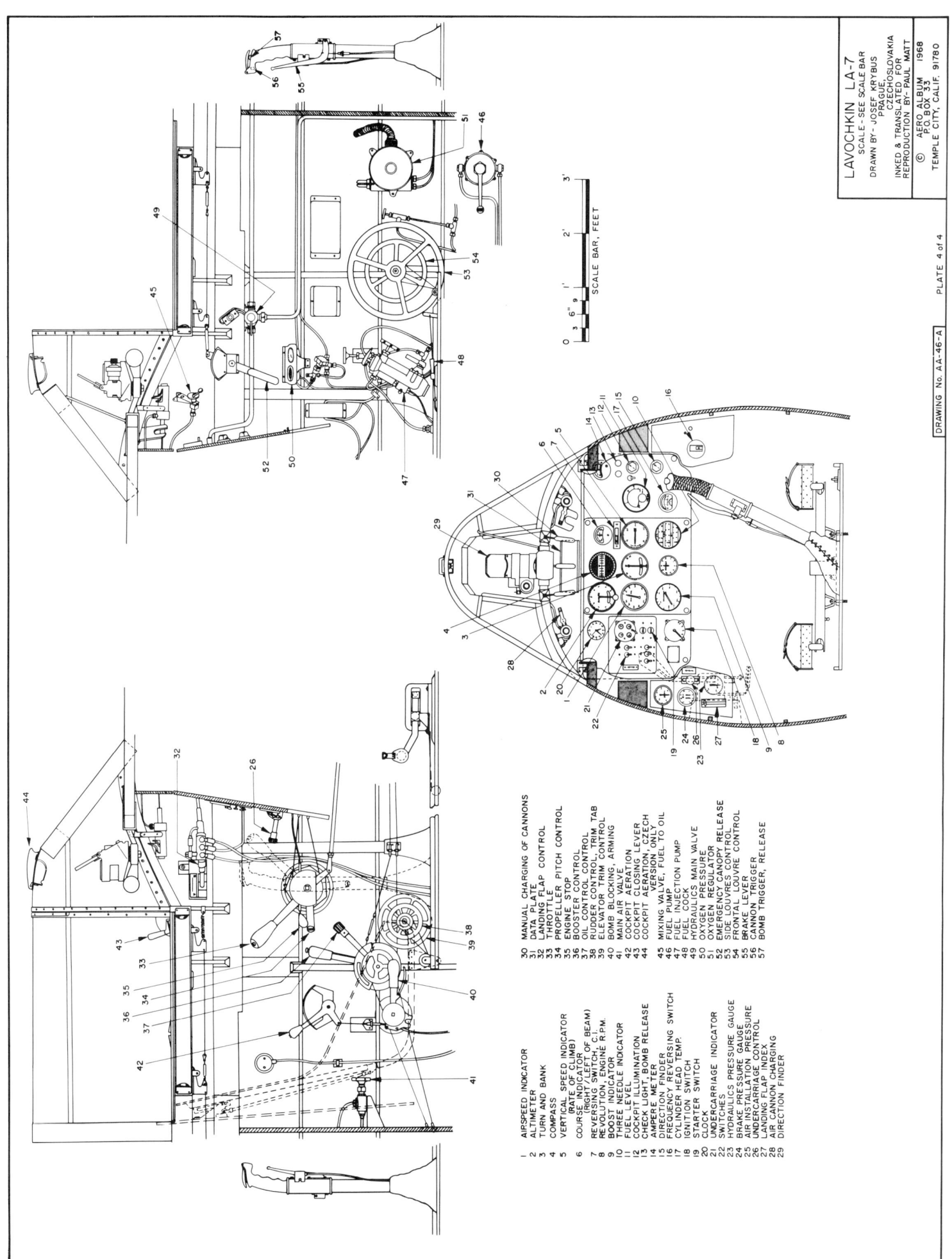

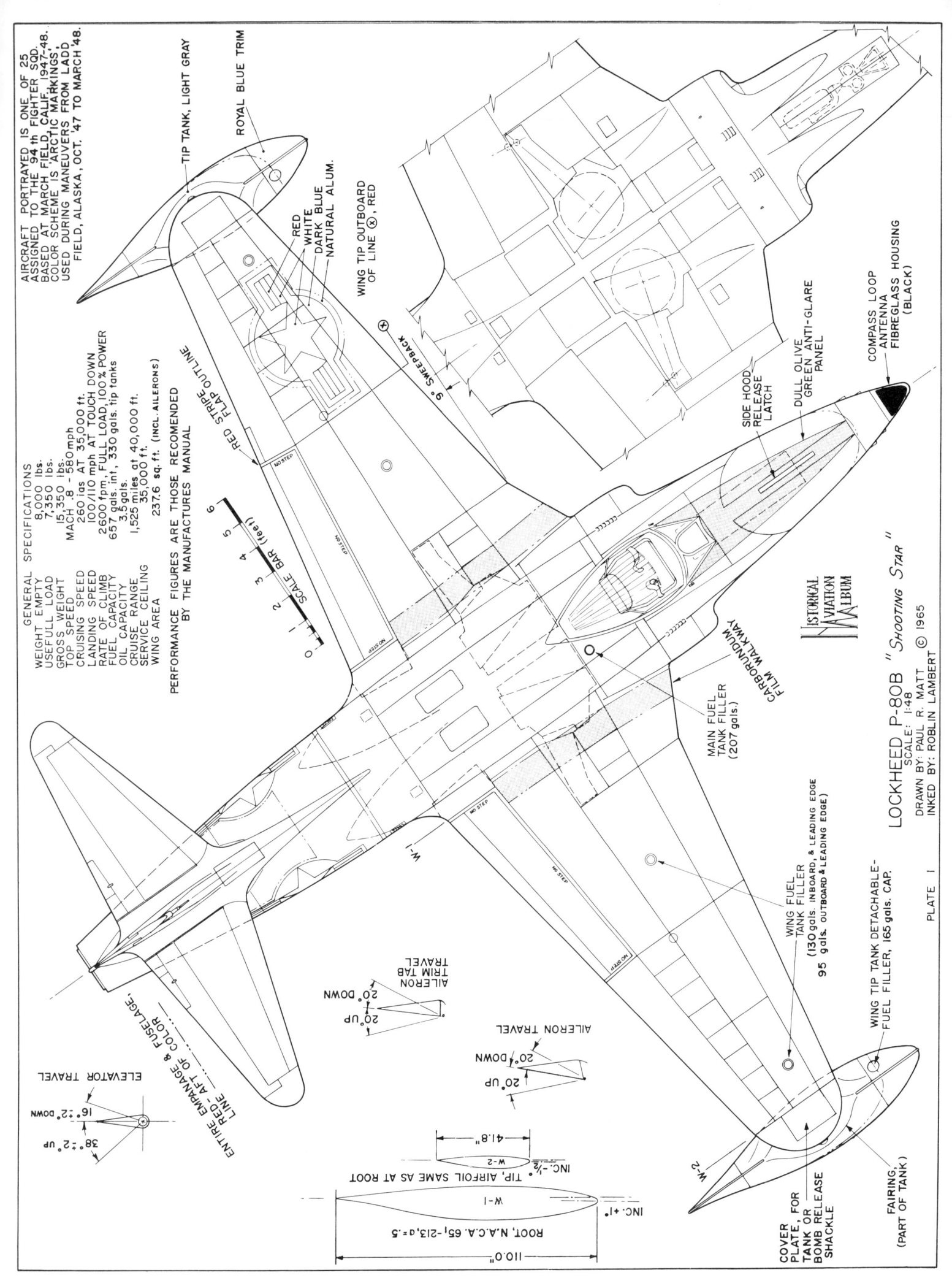

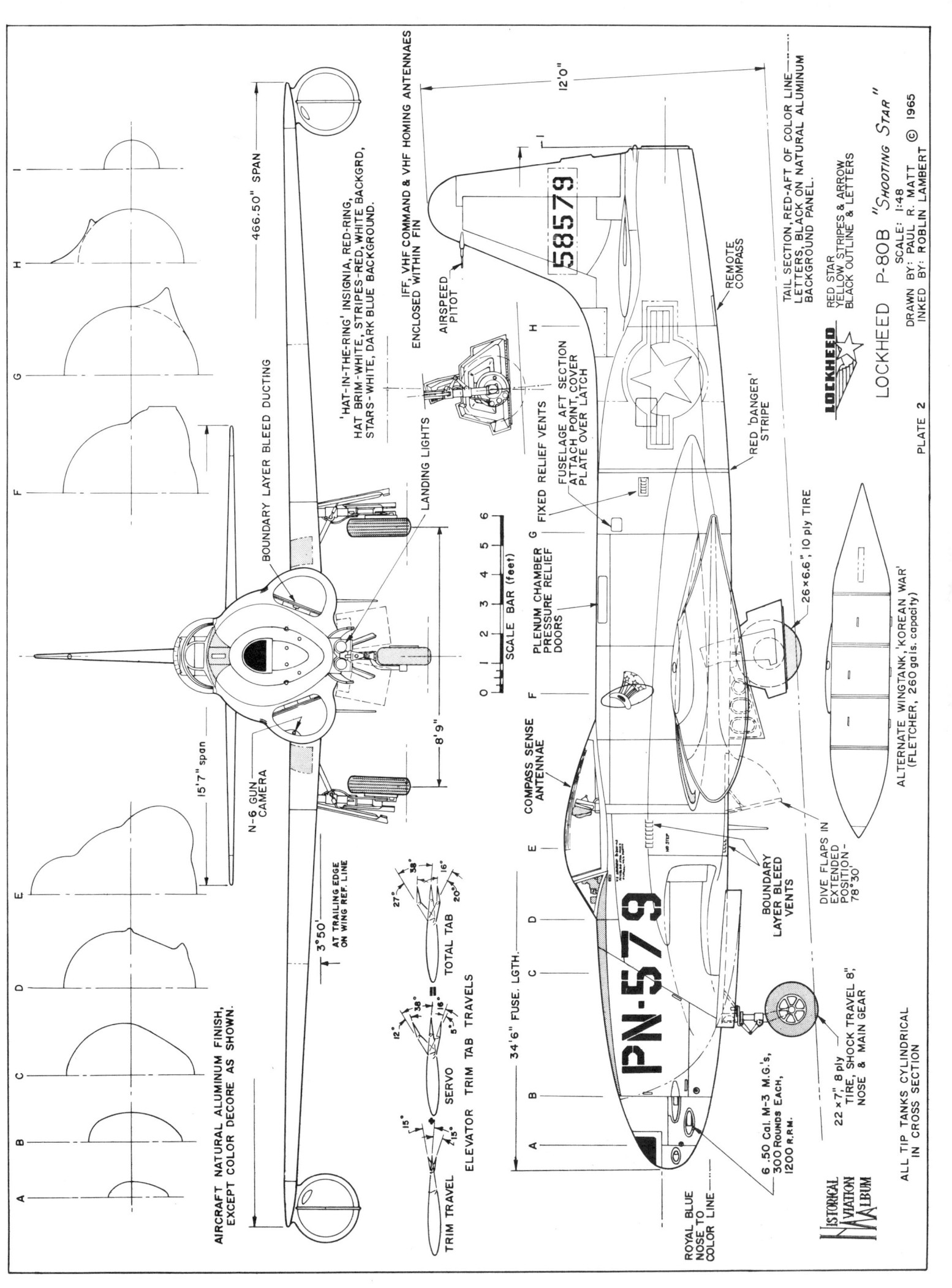

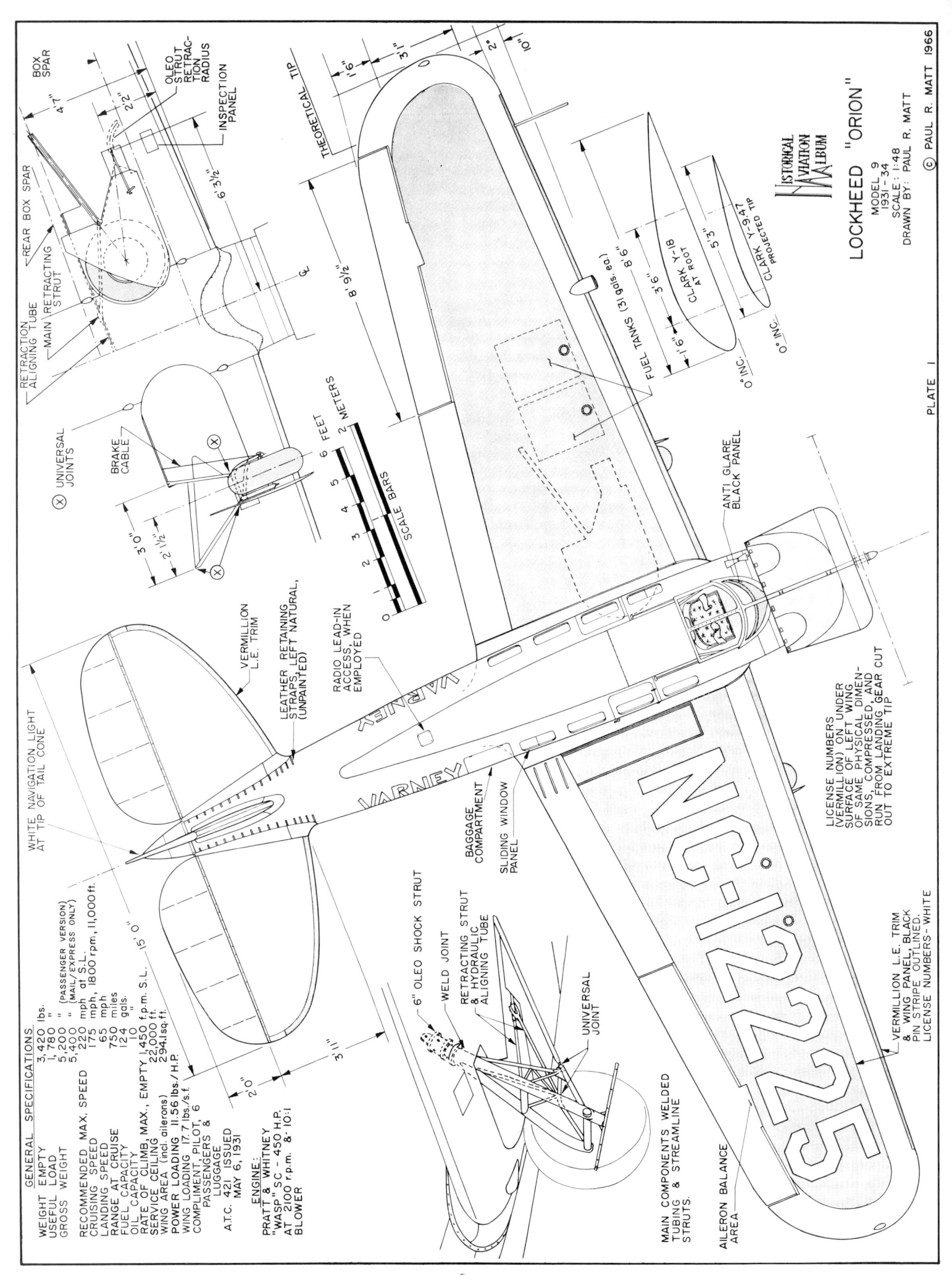

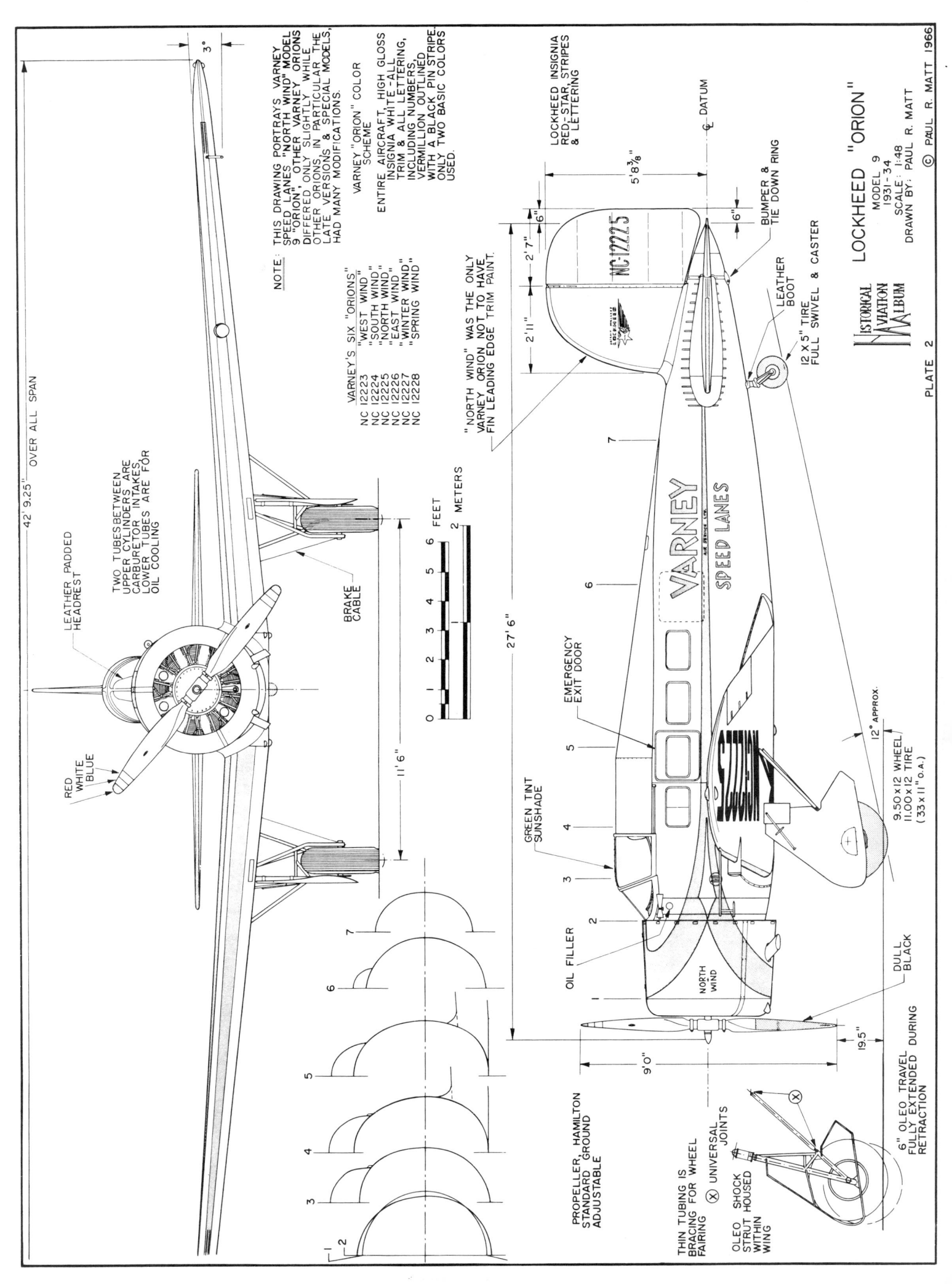

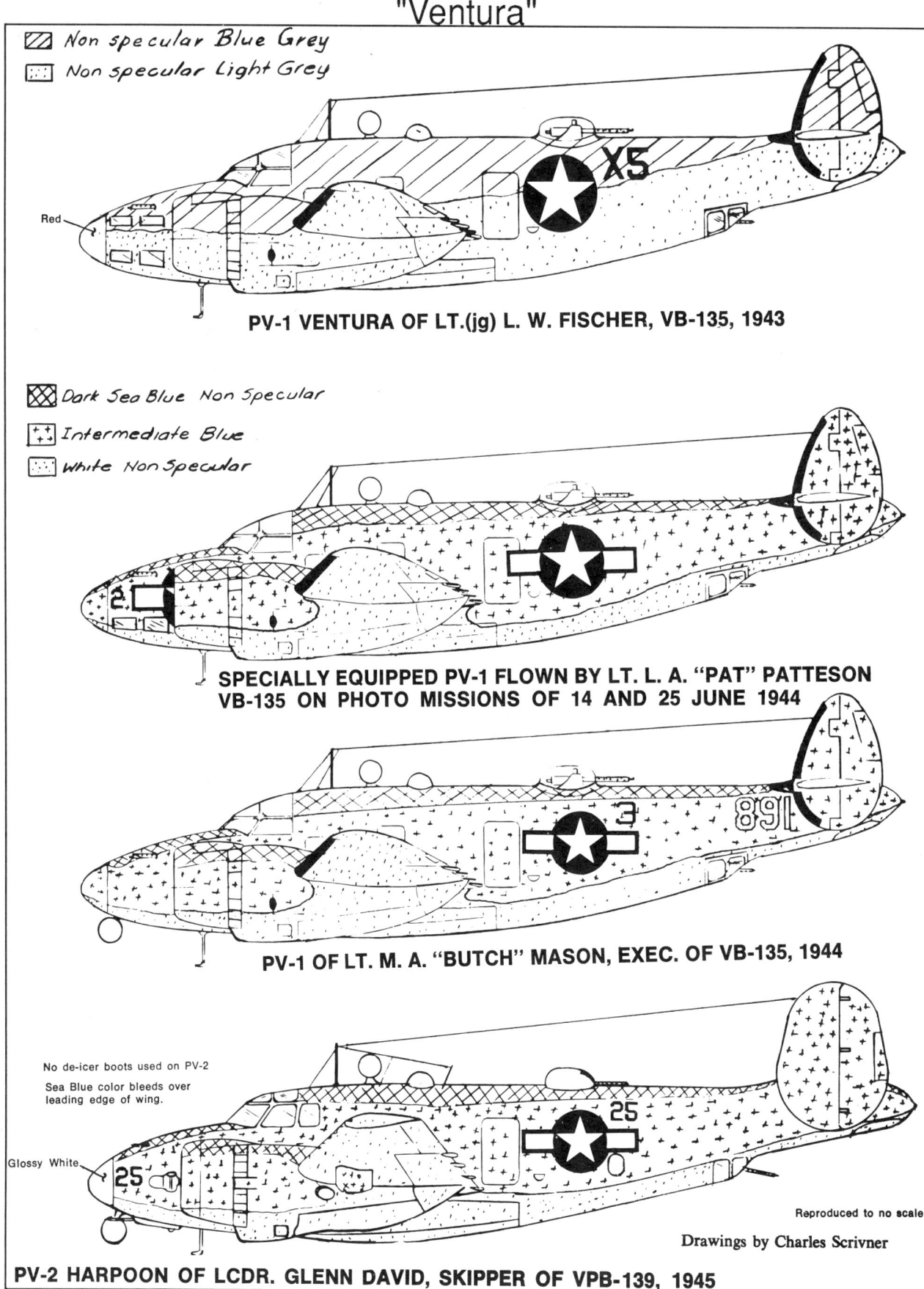

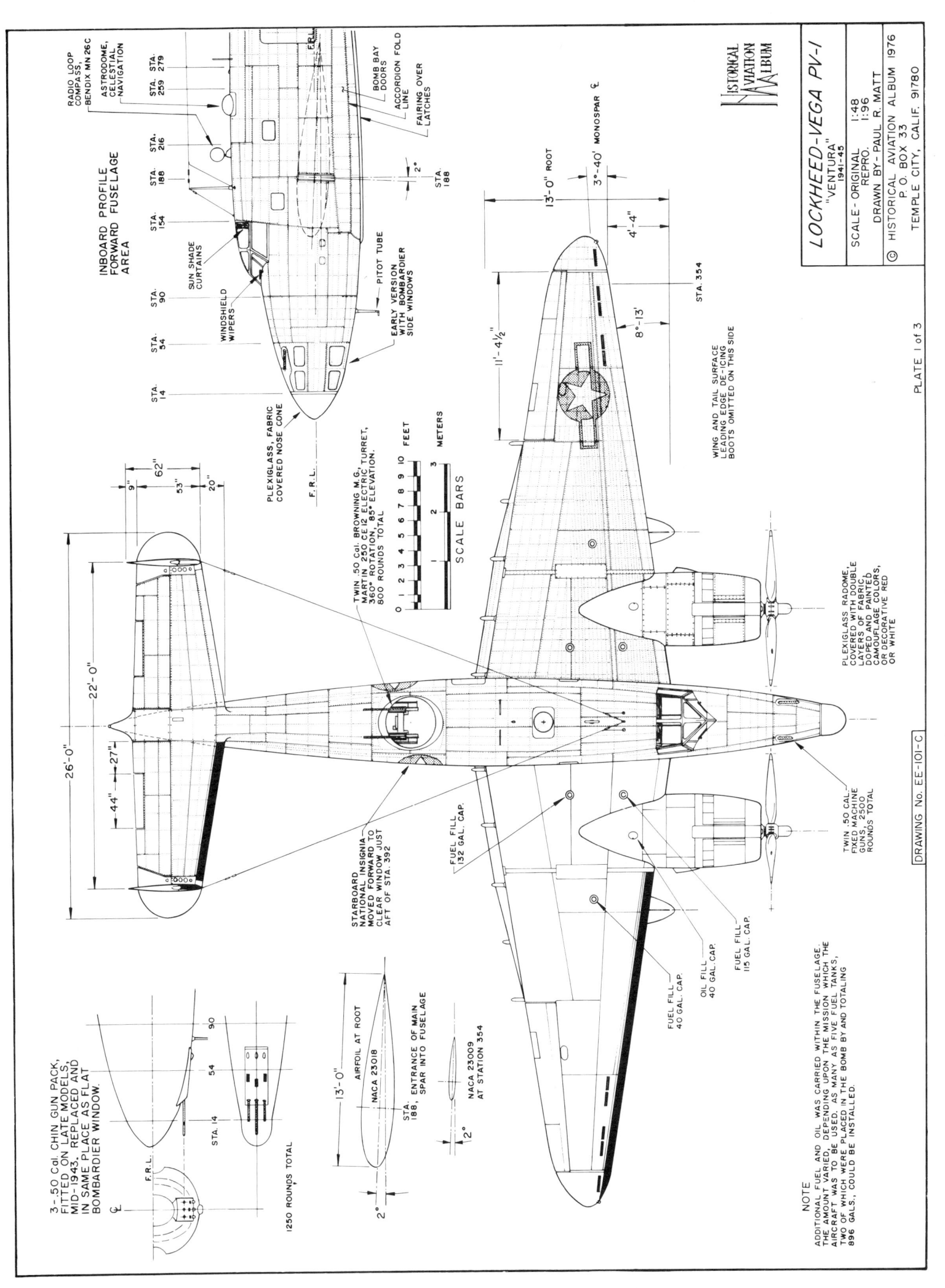

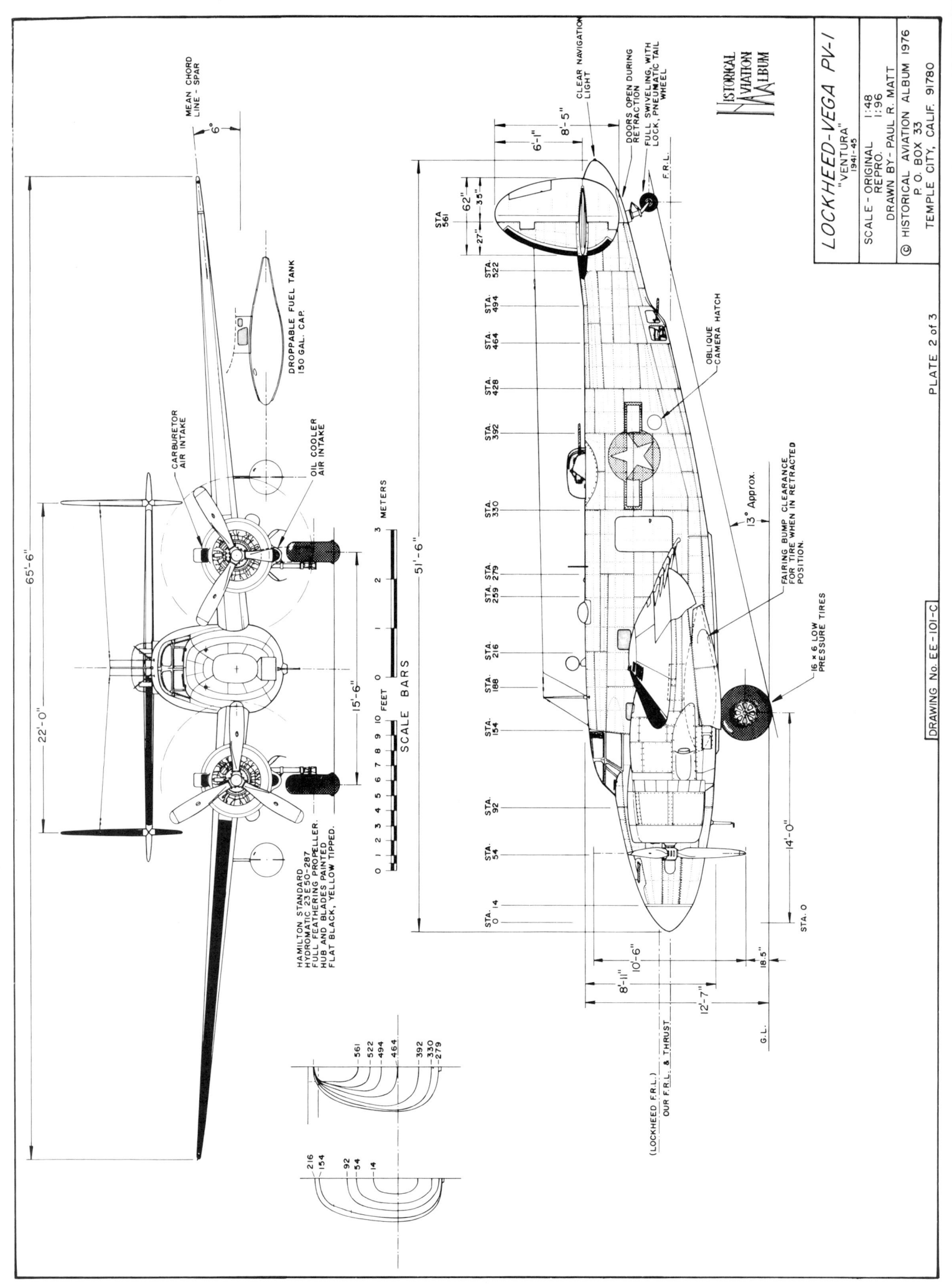

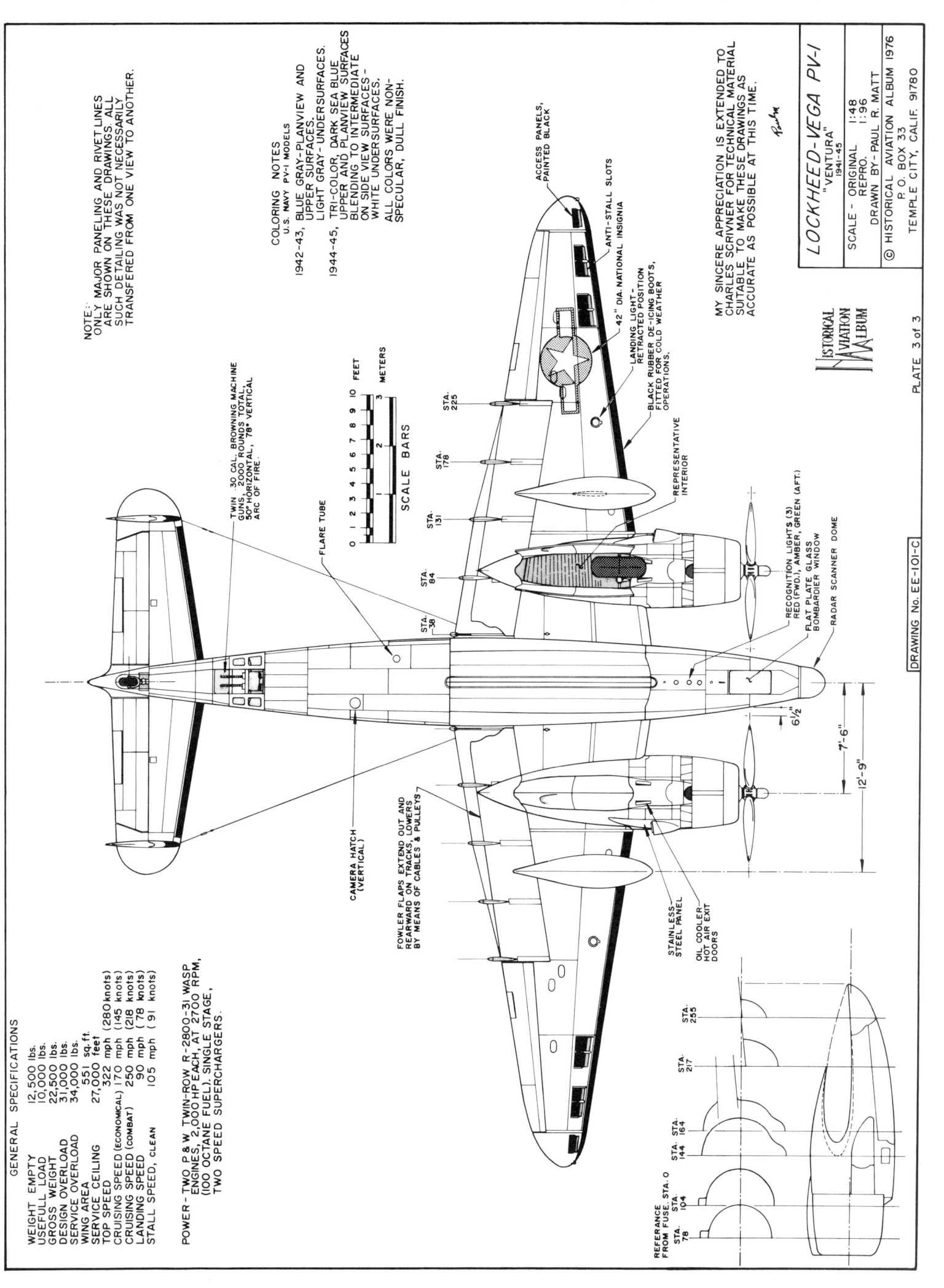

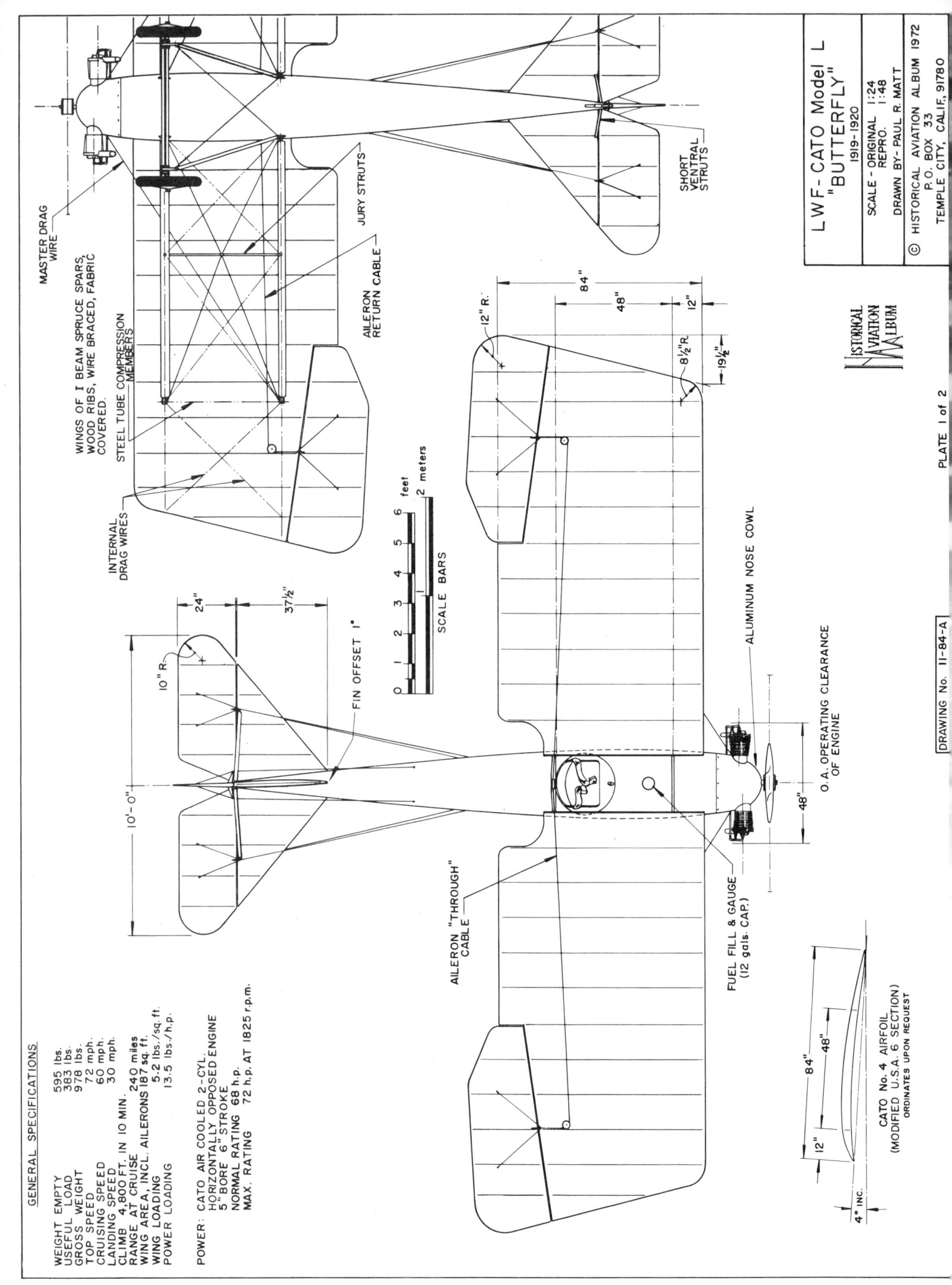

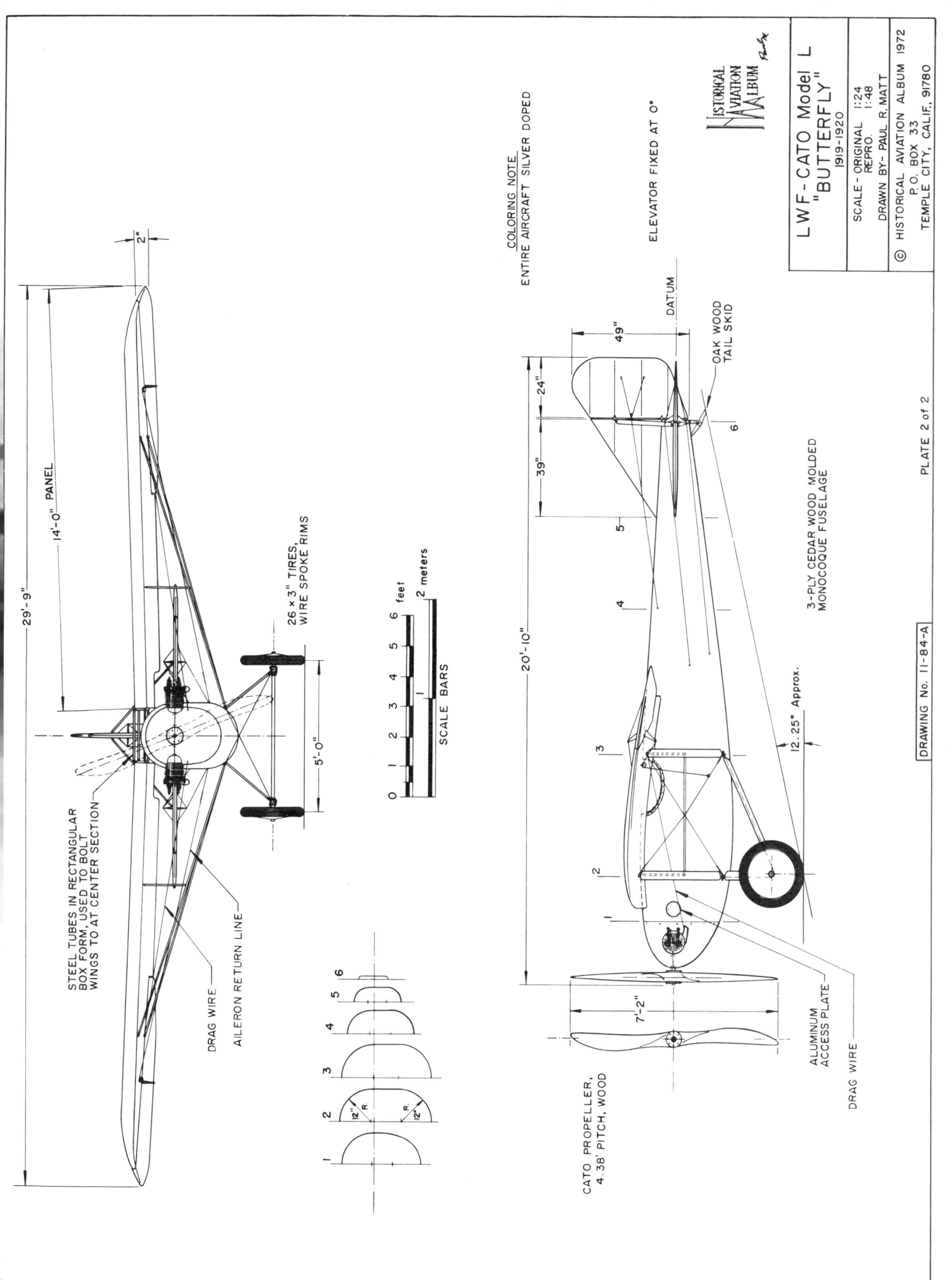

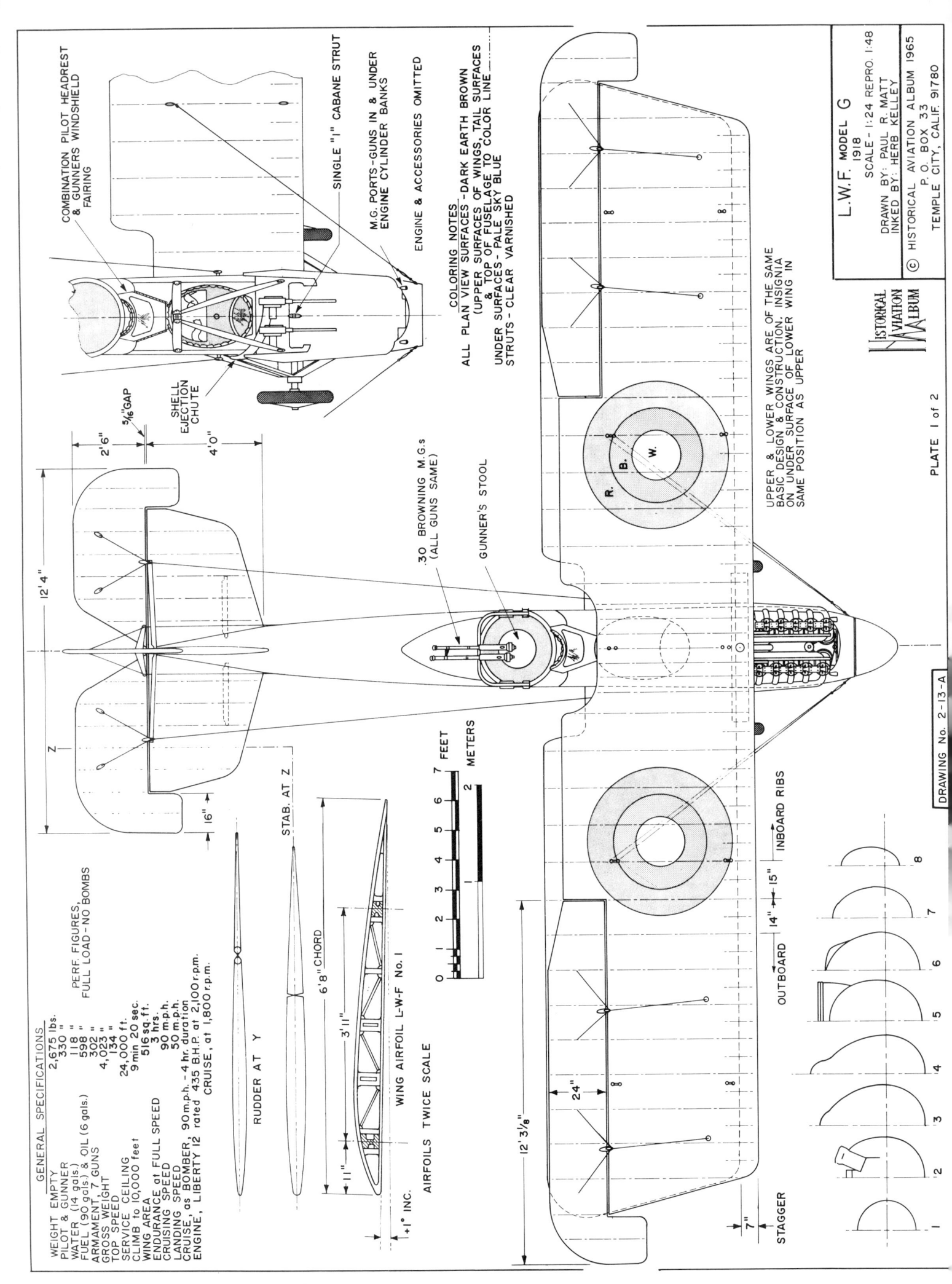

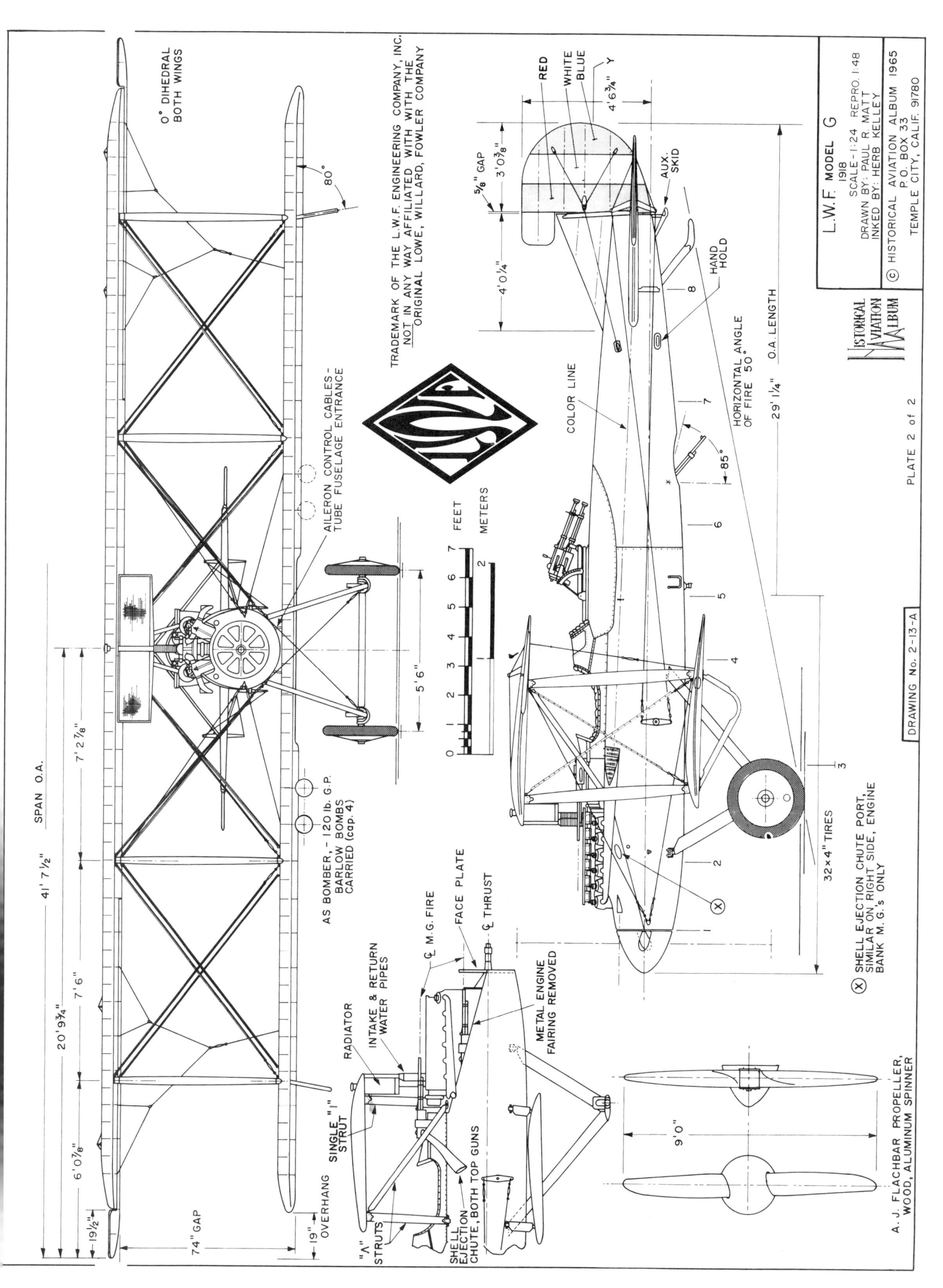

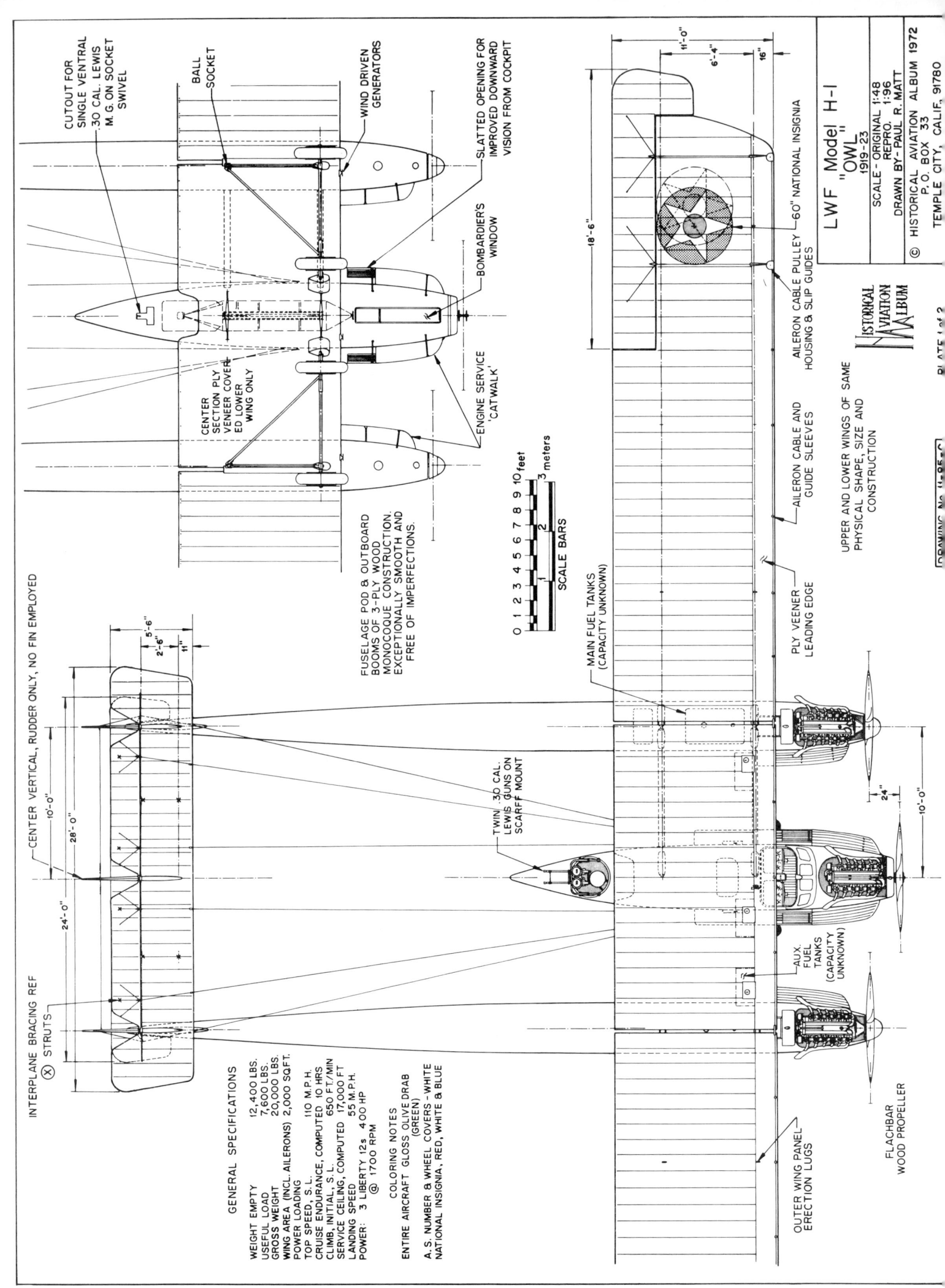

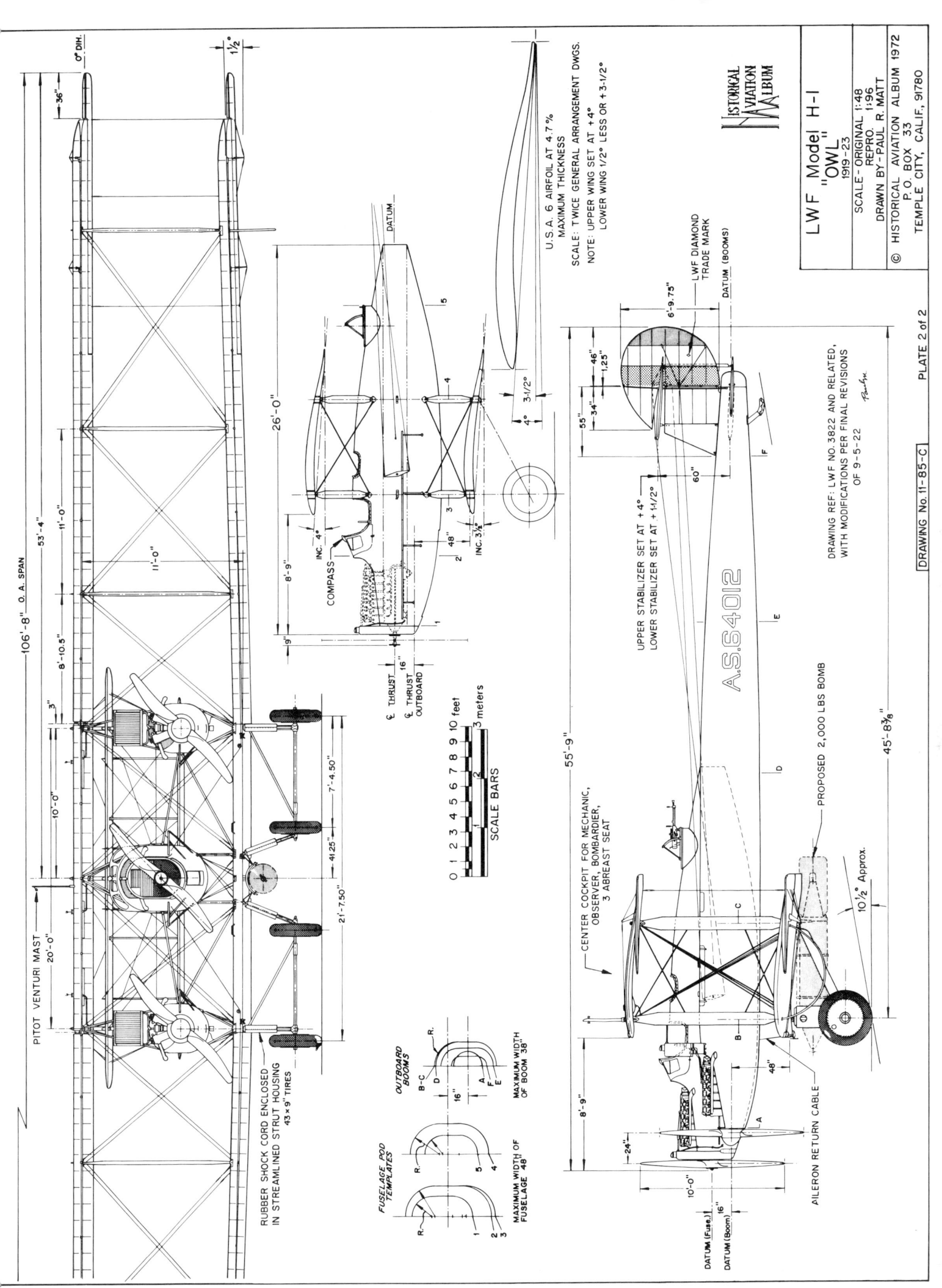

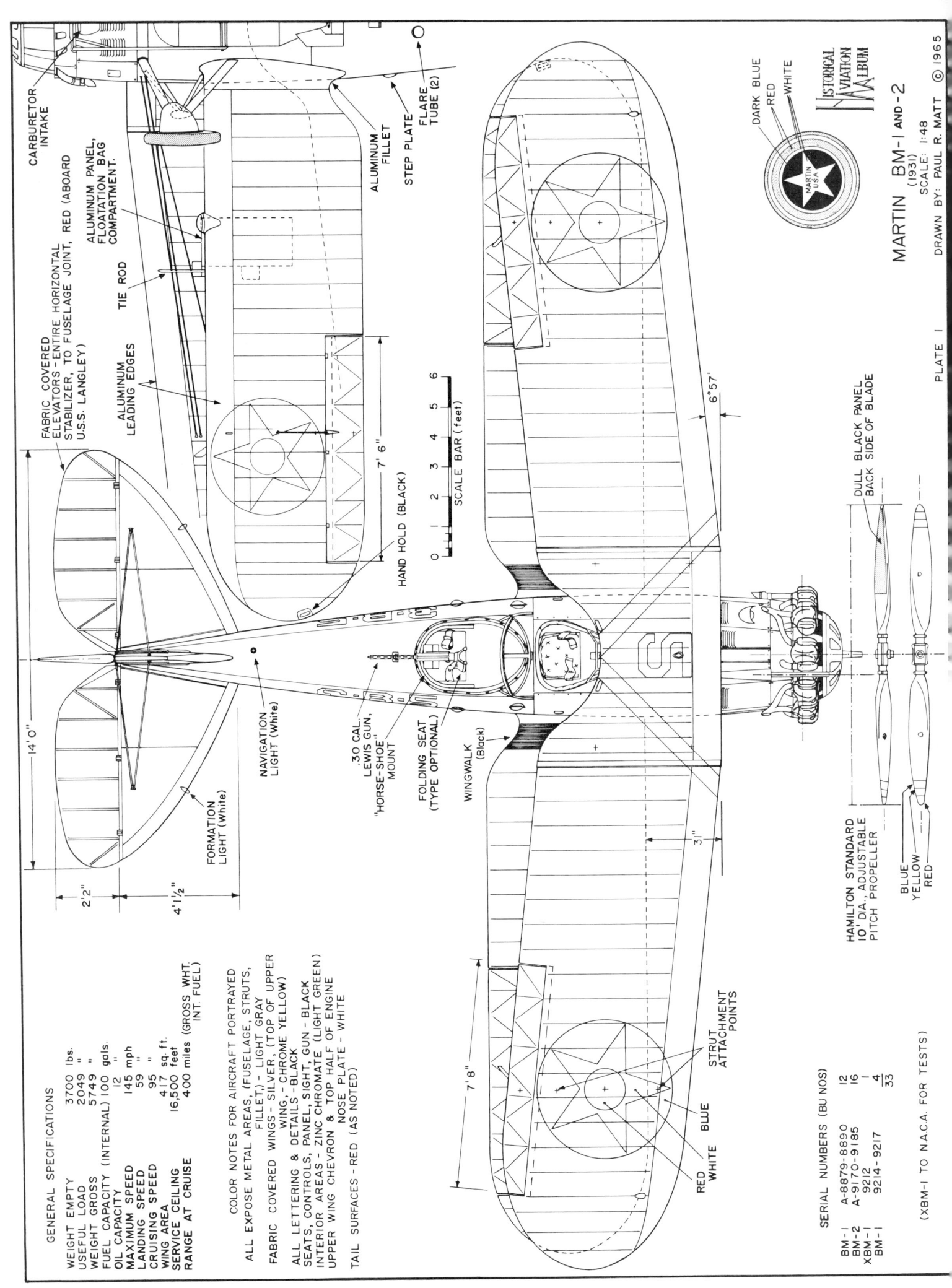

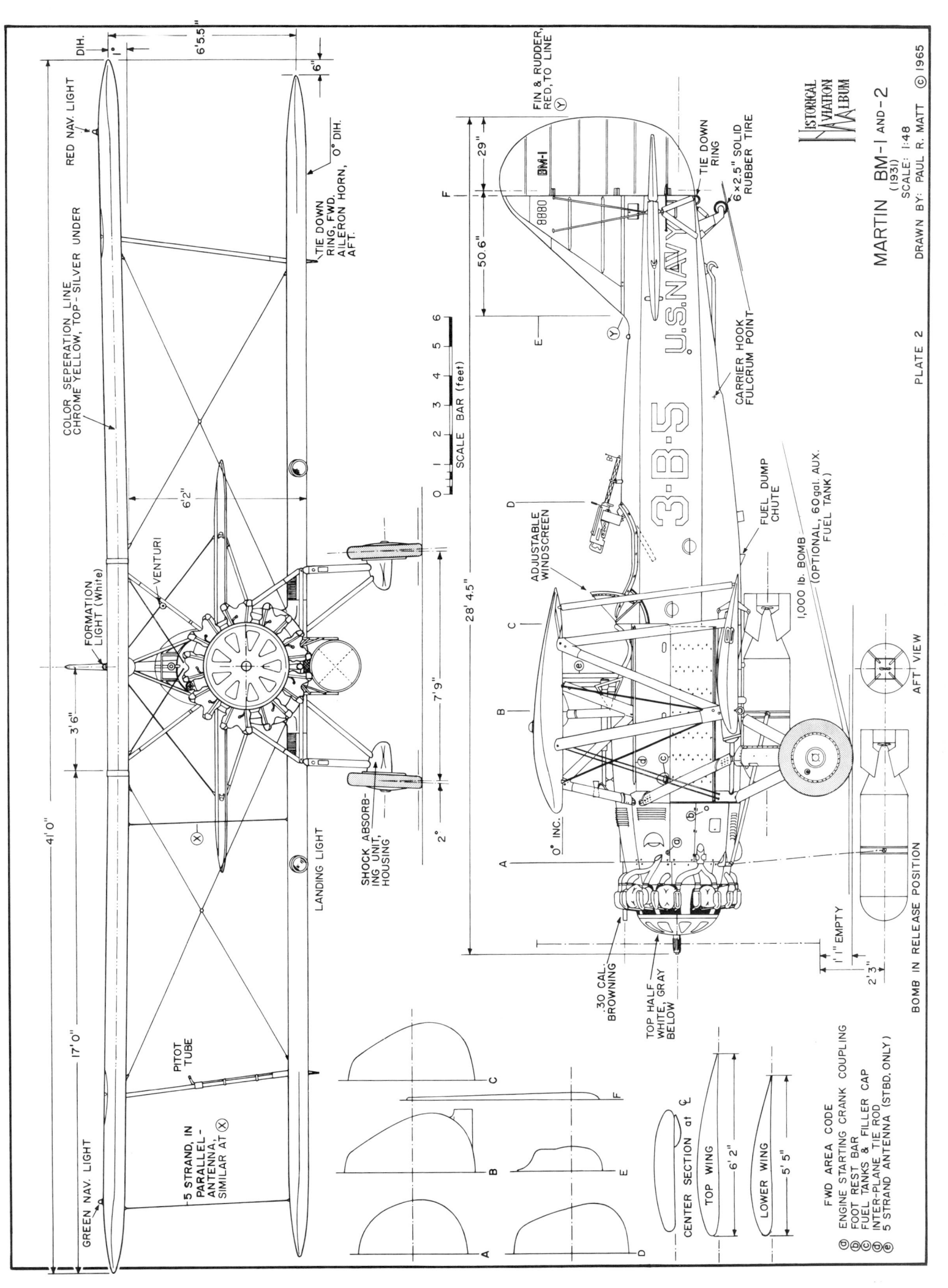

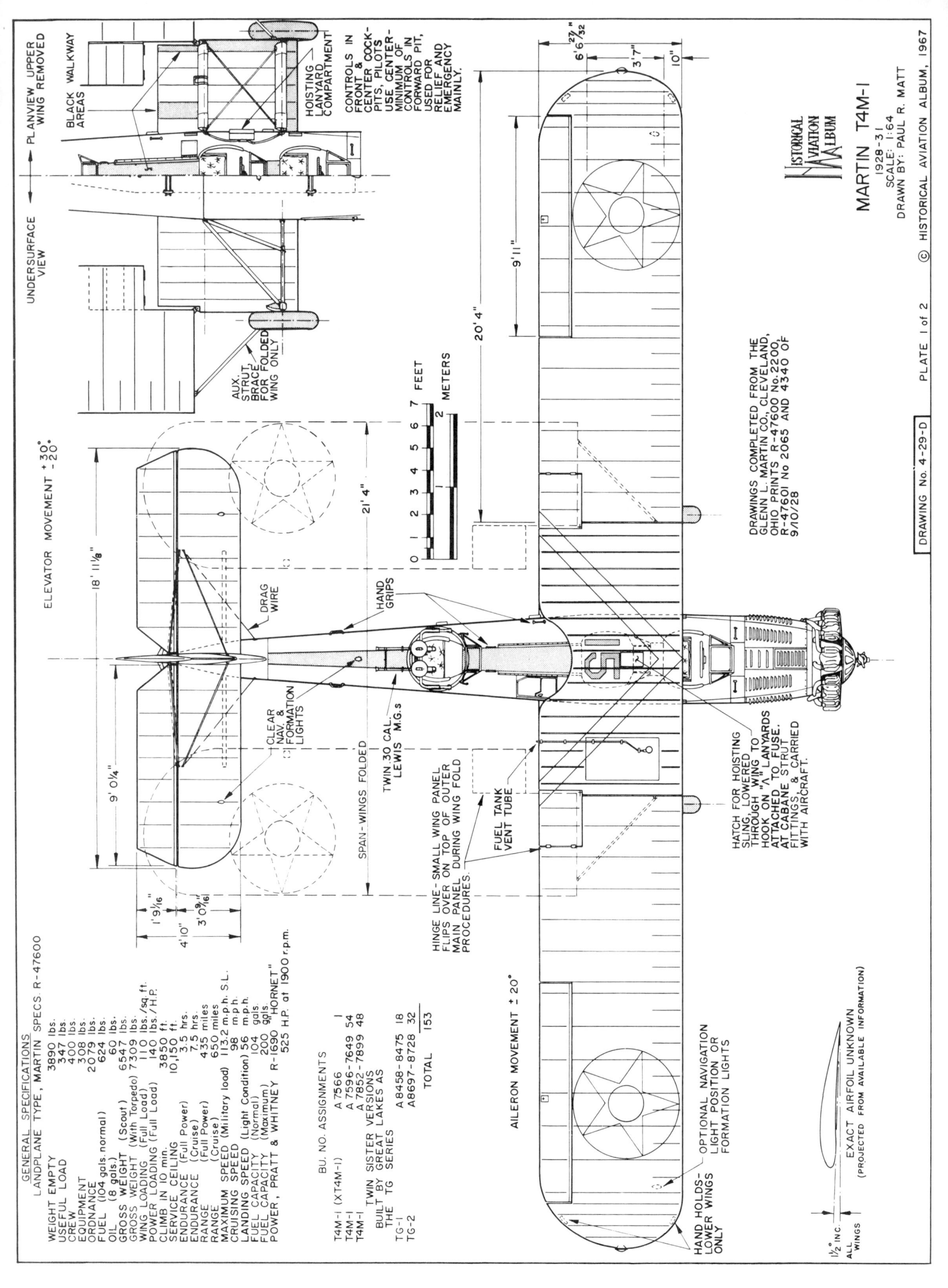

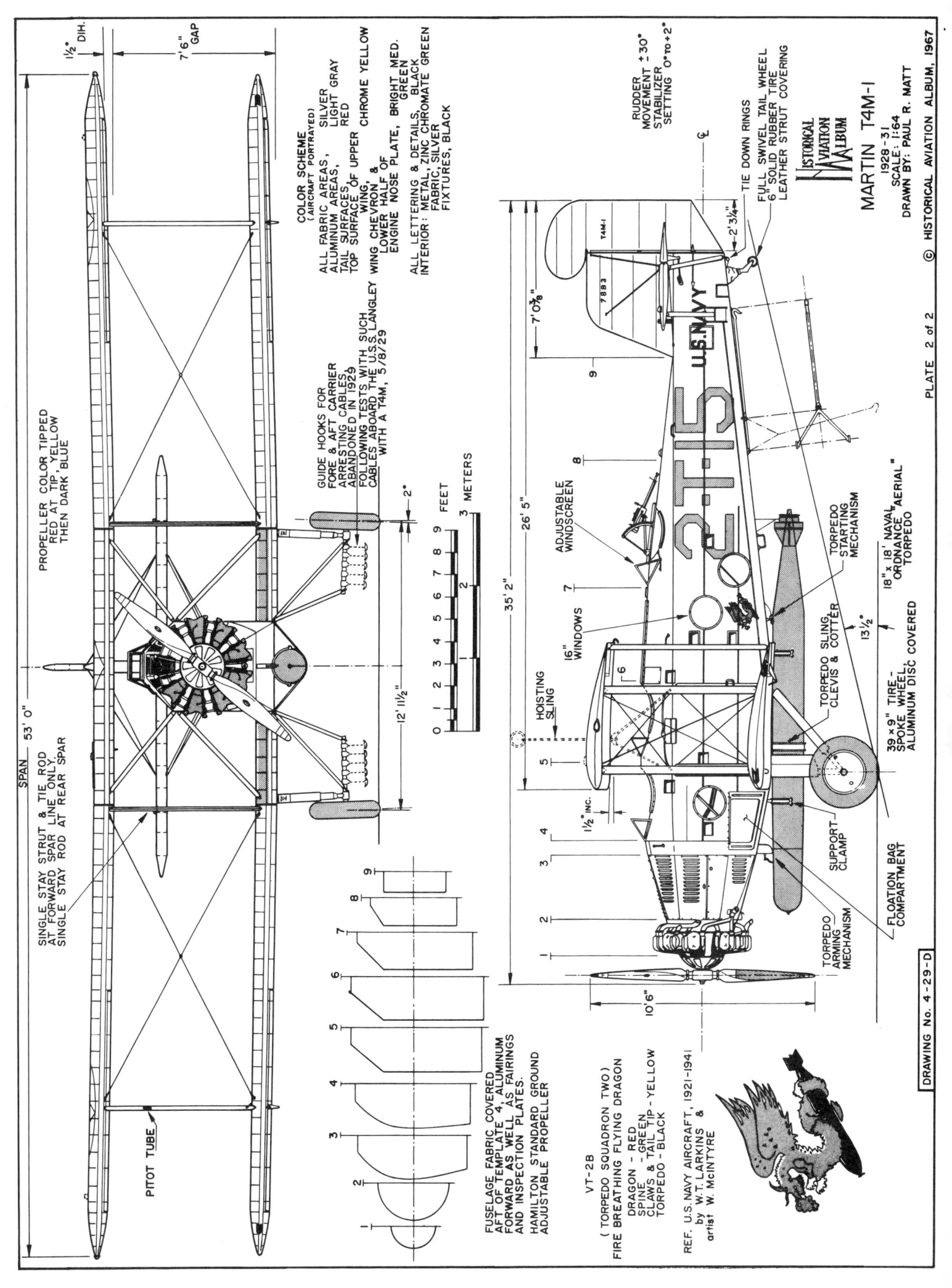

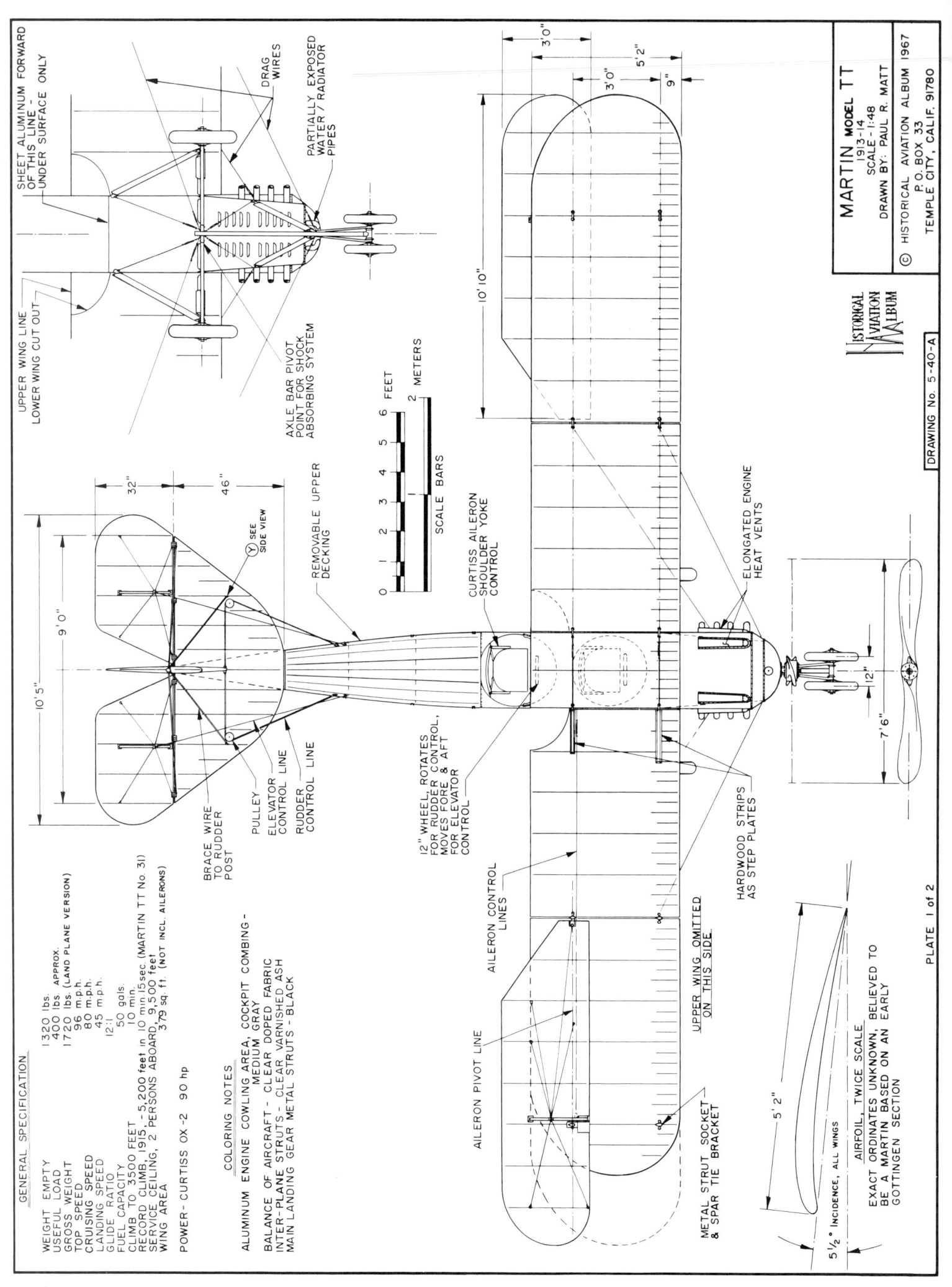

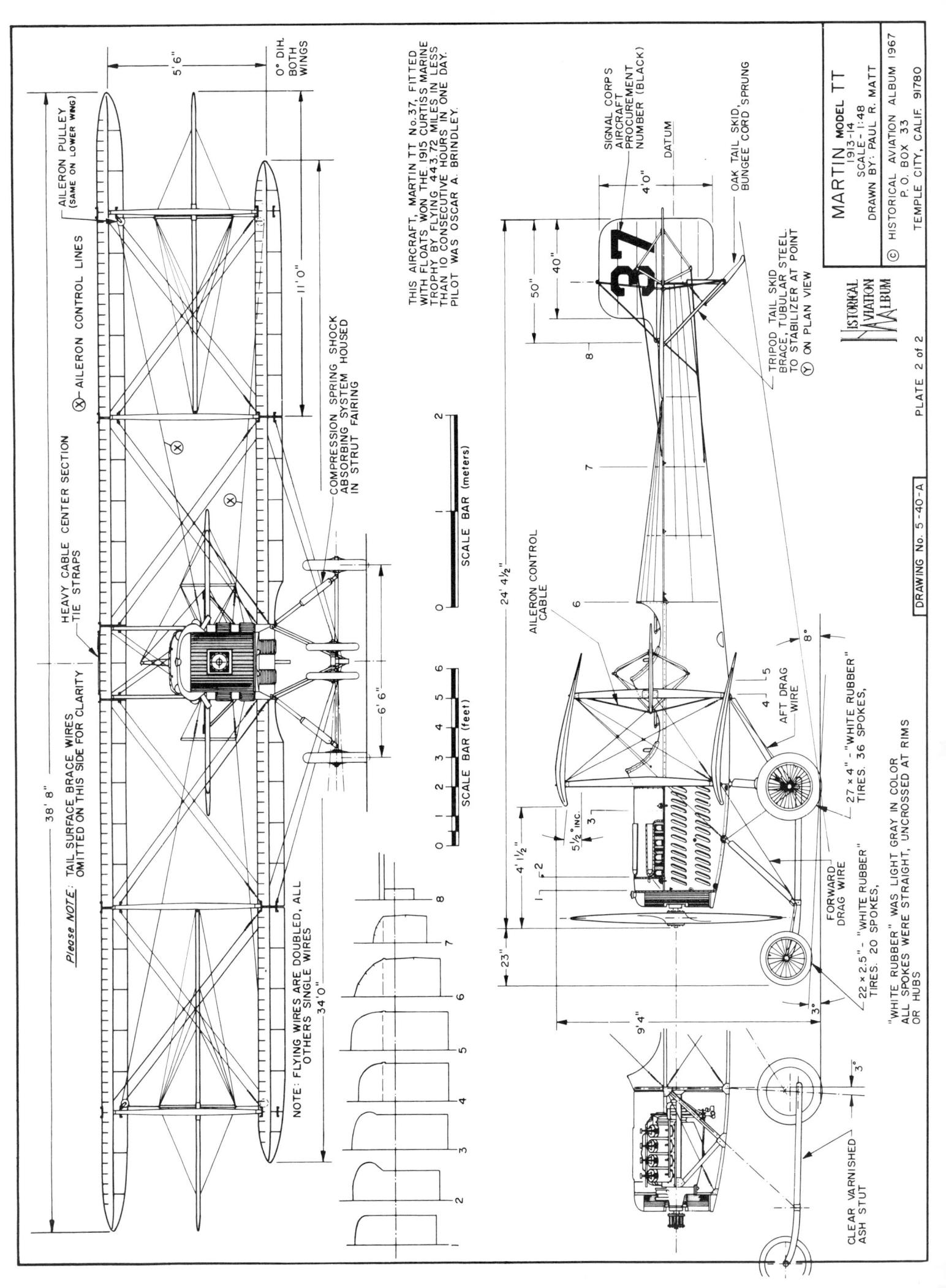

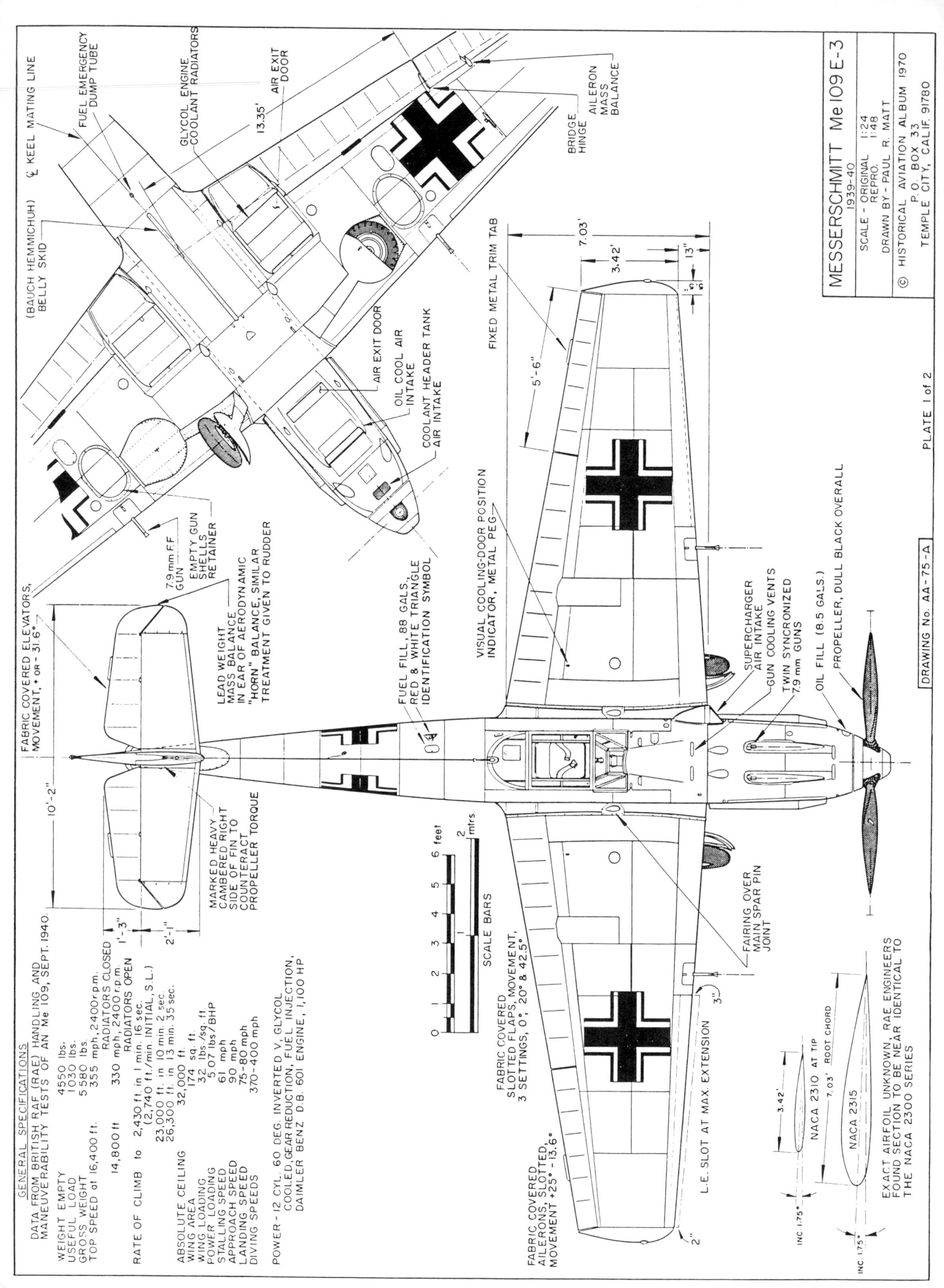

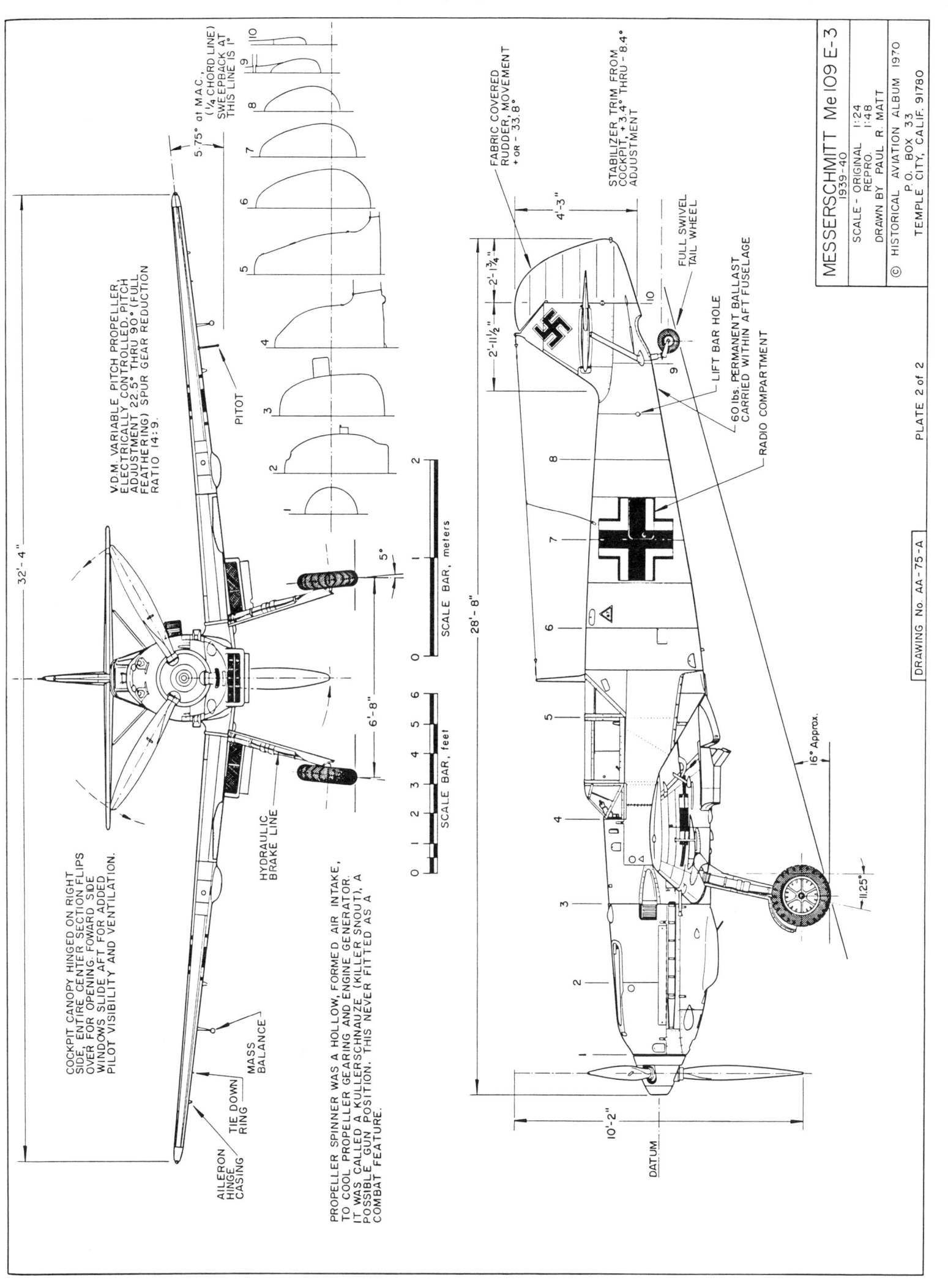

Morehouse 29.IH.P. Aircraft Engine
1925

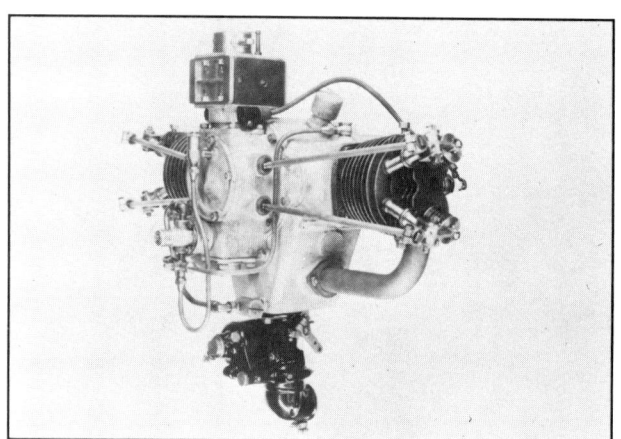

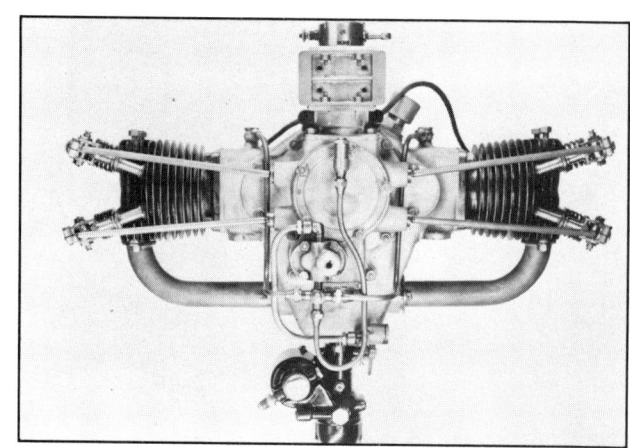

Top left: Front view

Top right: Top view

Middle left: Right rear view

Middle right: Direct rear view

Bottom right: Bottom view

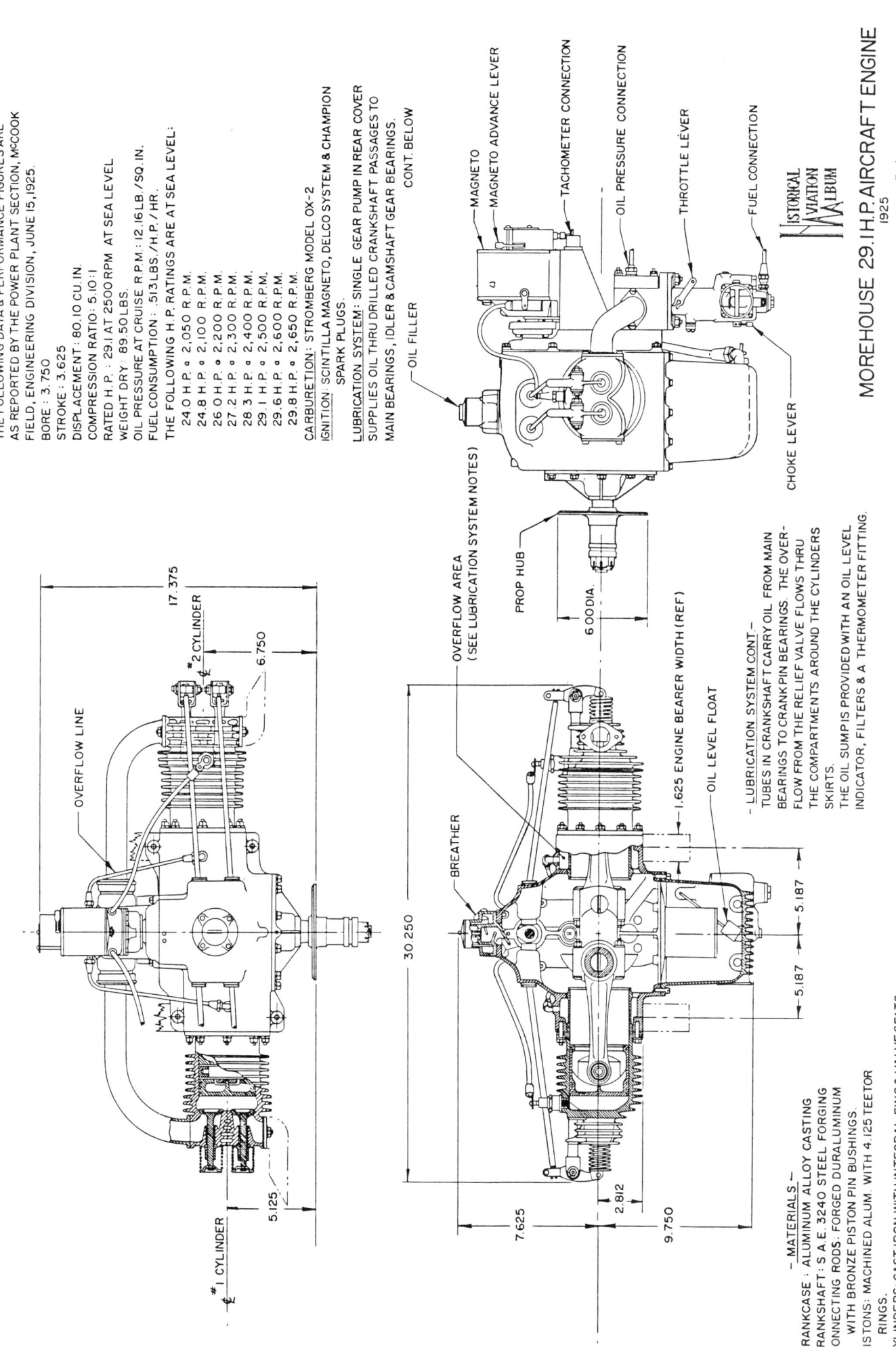

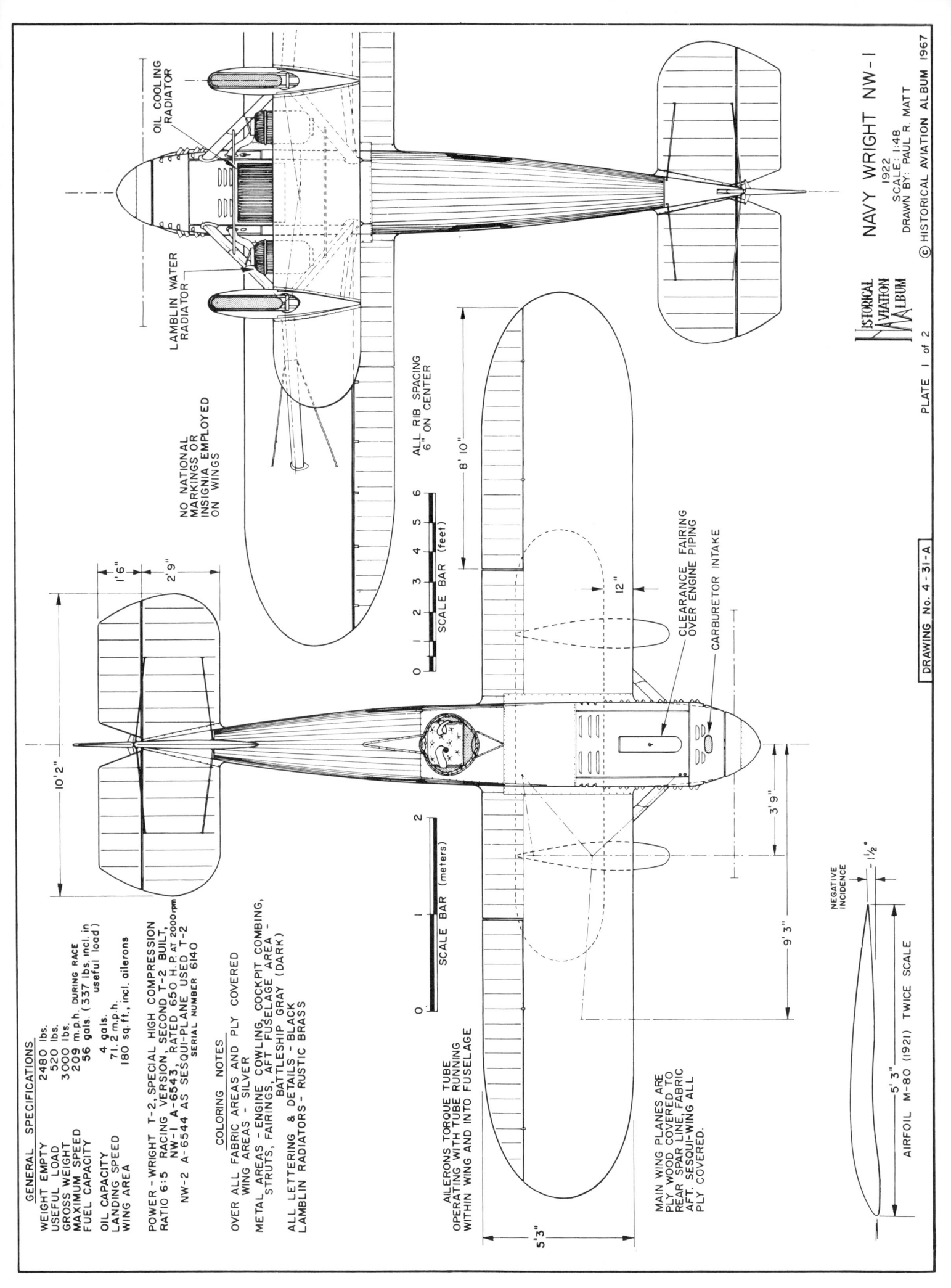

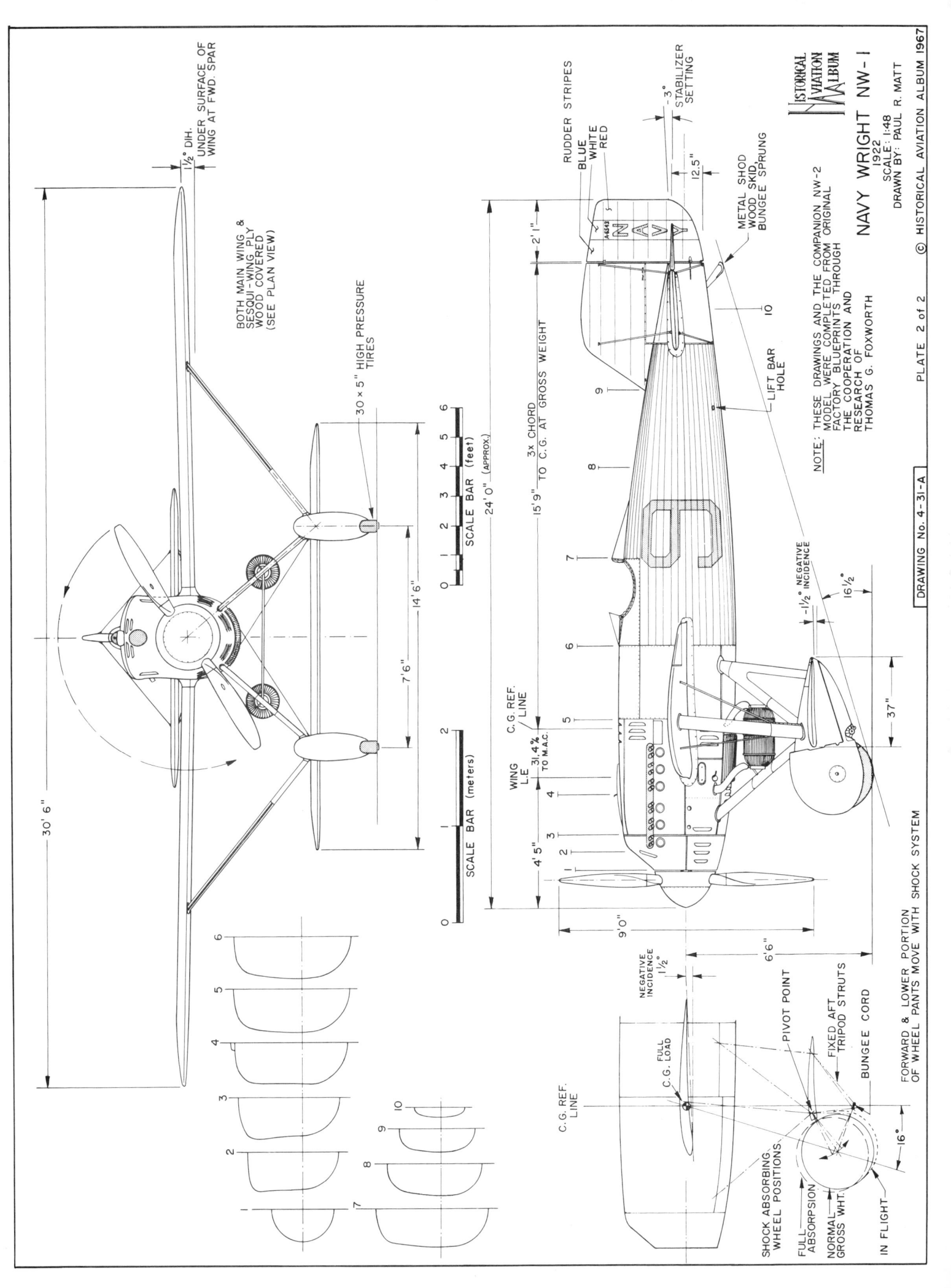

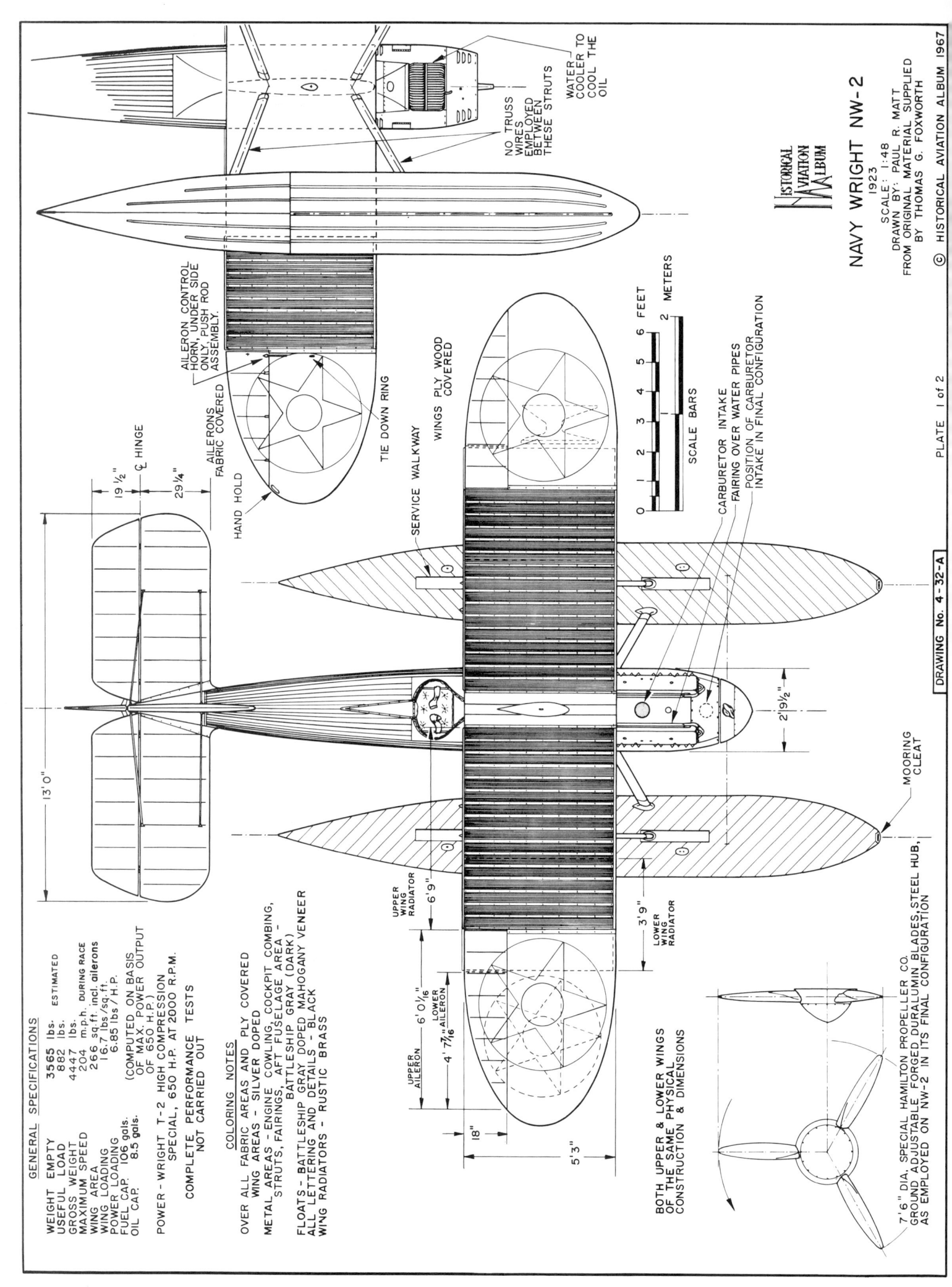

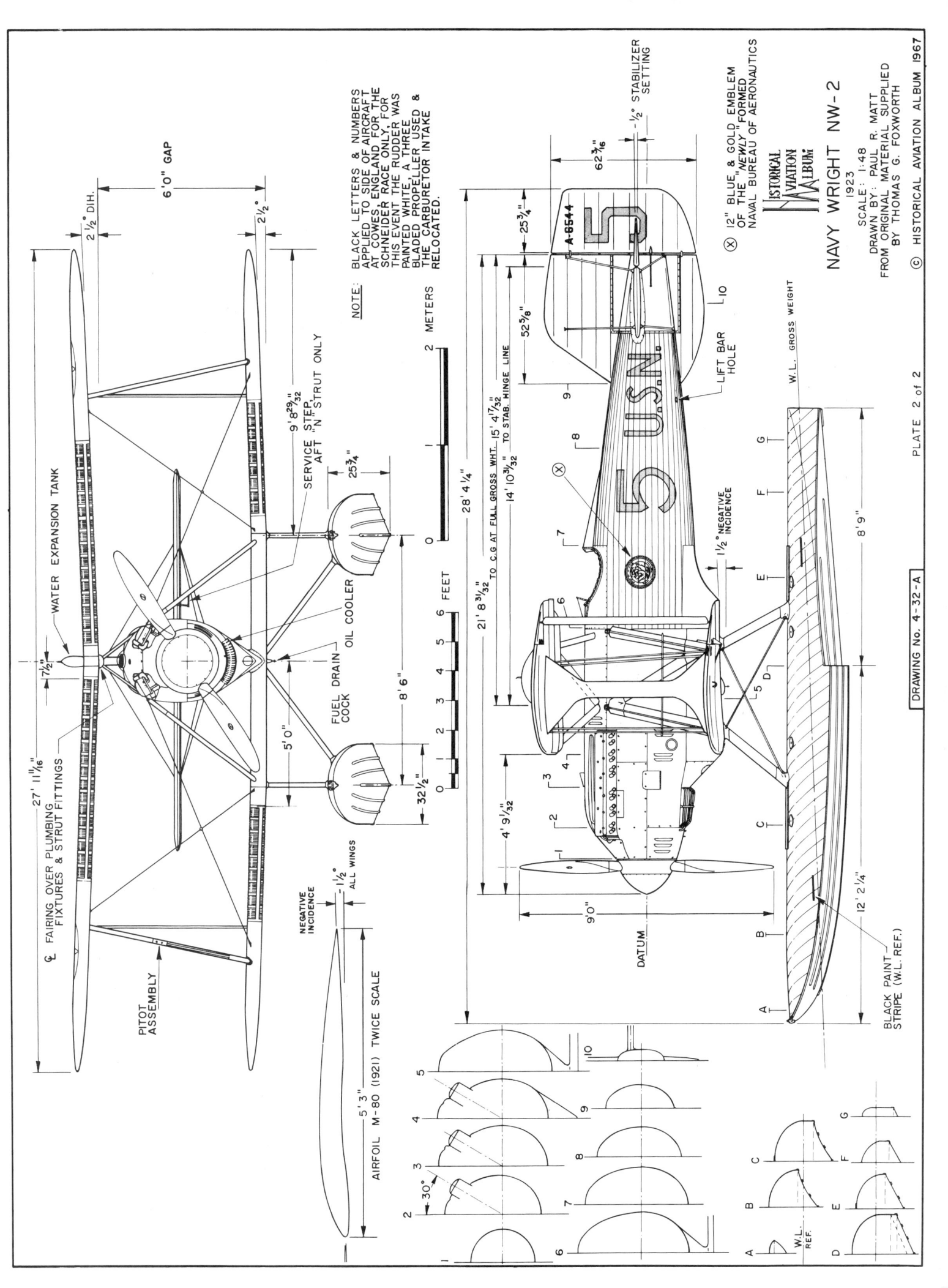

North American O-47A

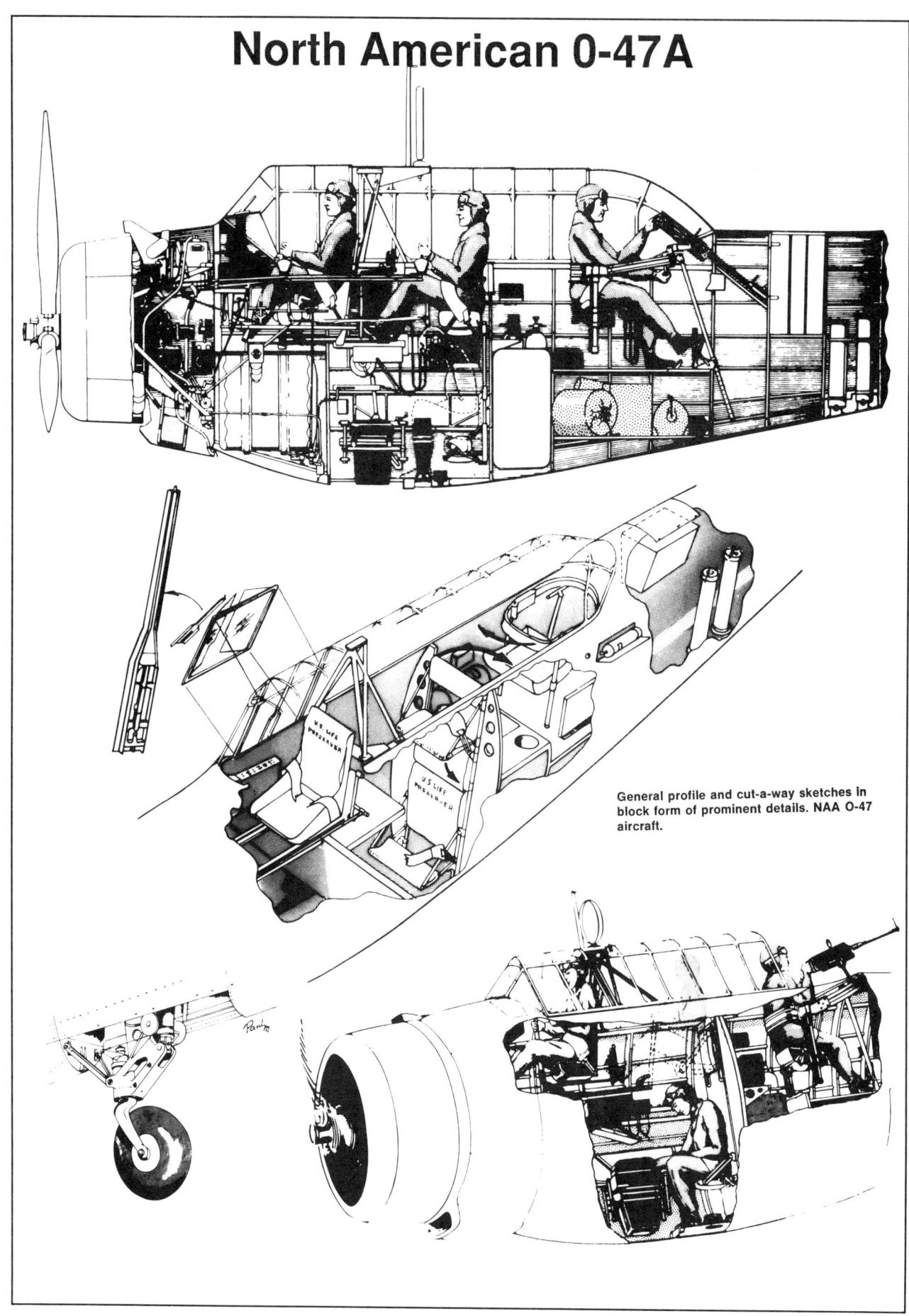

General profile and cut-a-way sketches in block form of prominent details. NAA O-47 aircraft.

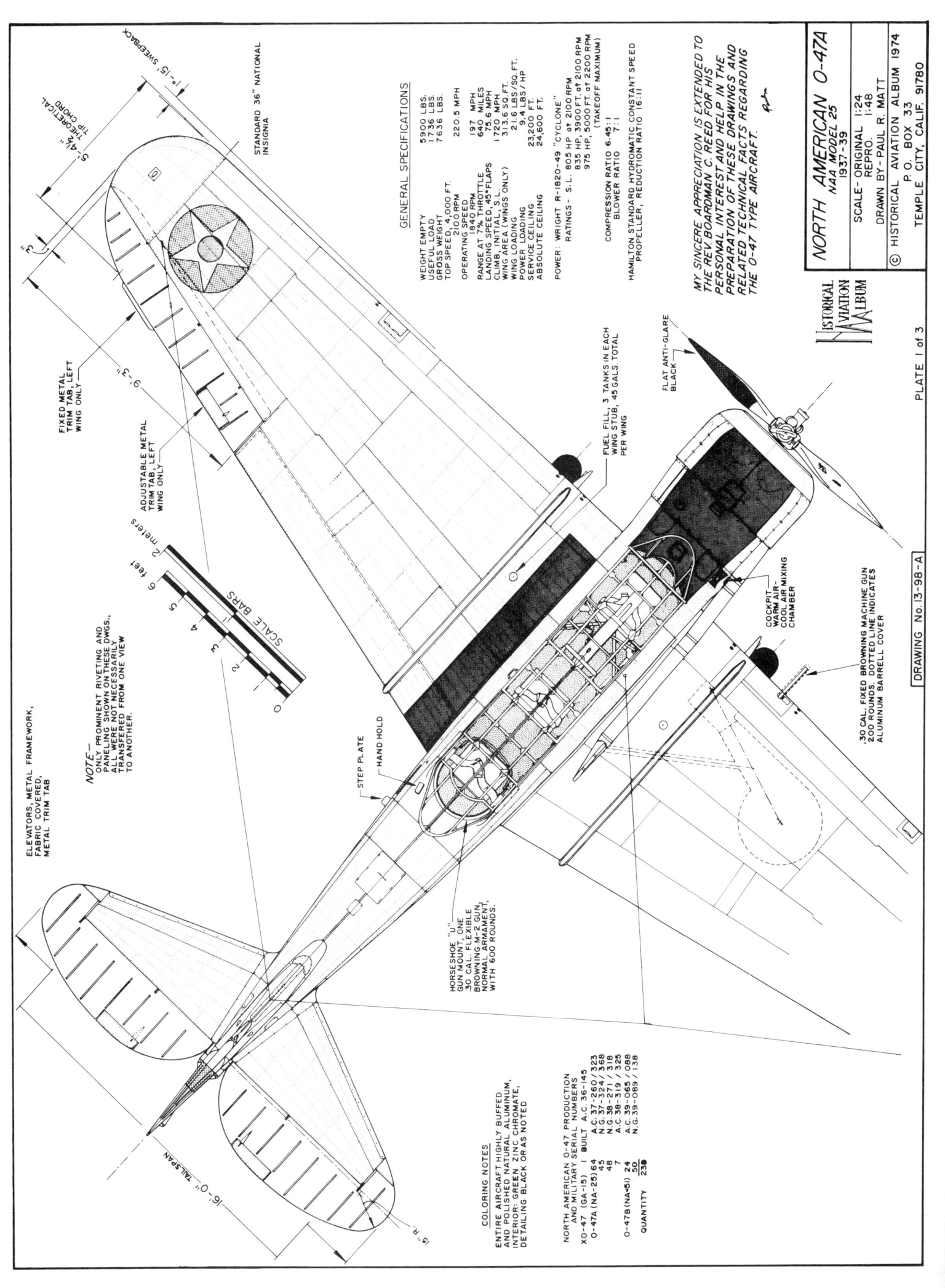

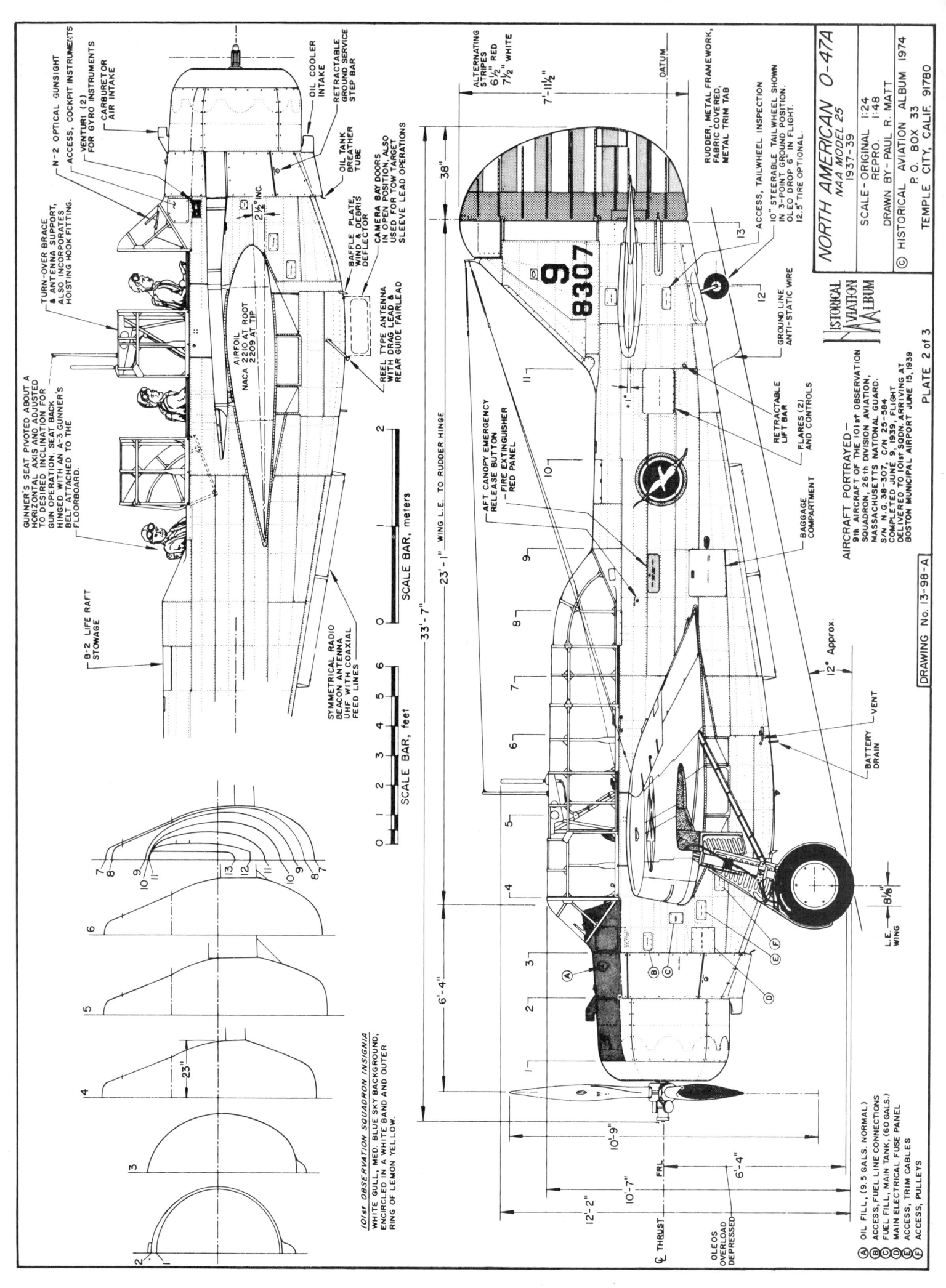

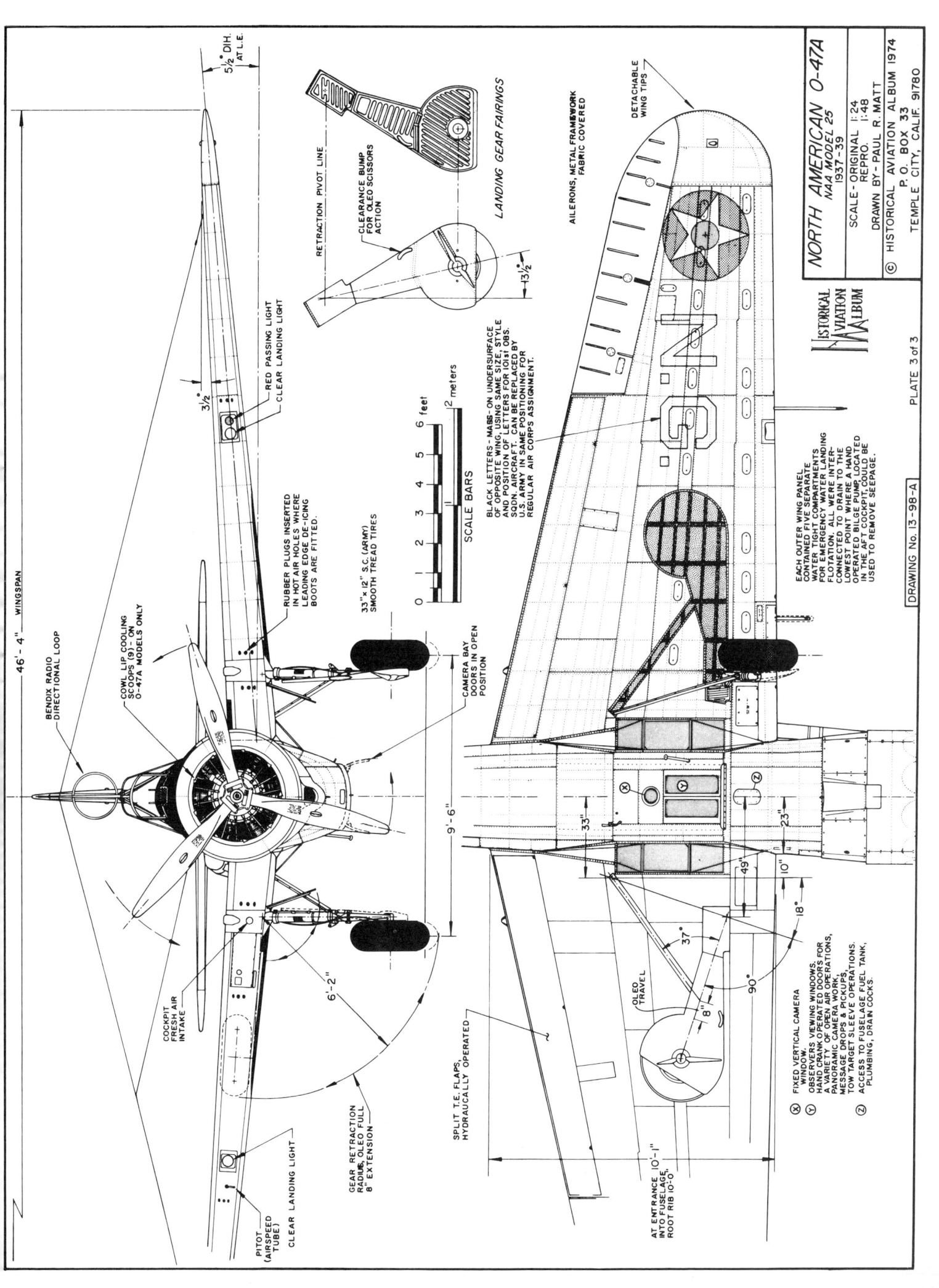

North American AT-6D "Texan"

North American AT-6D, September 6, 1948.

North American BT-9C. Front cockpit.

North American AT-6. Rear cockpit.

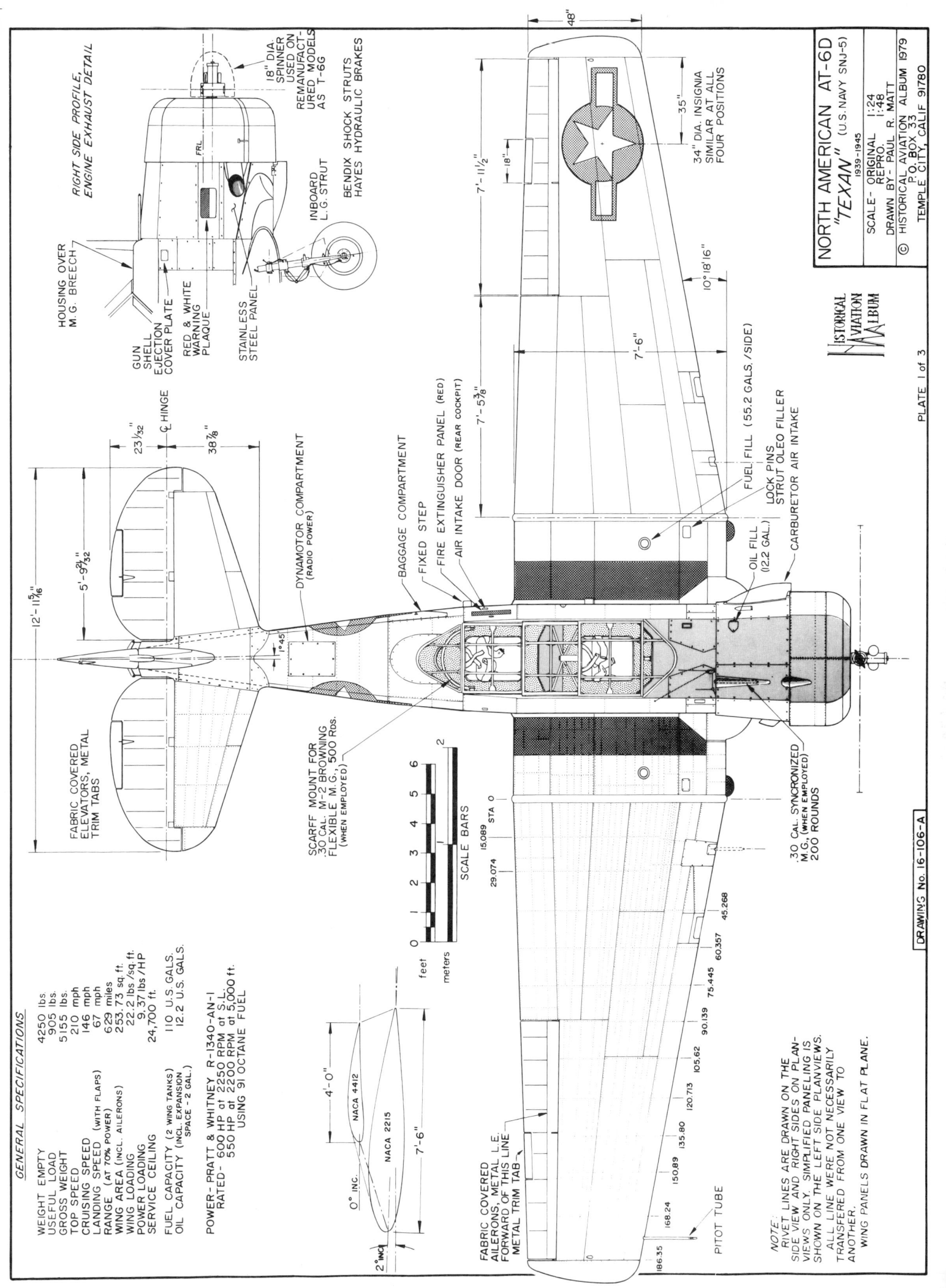

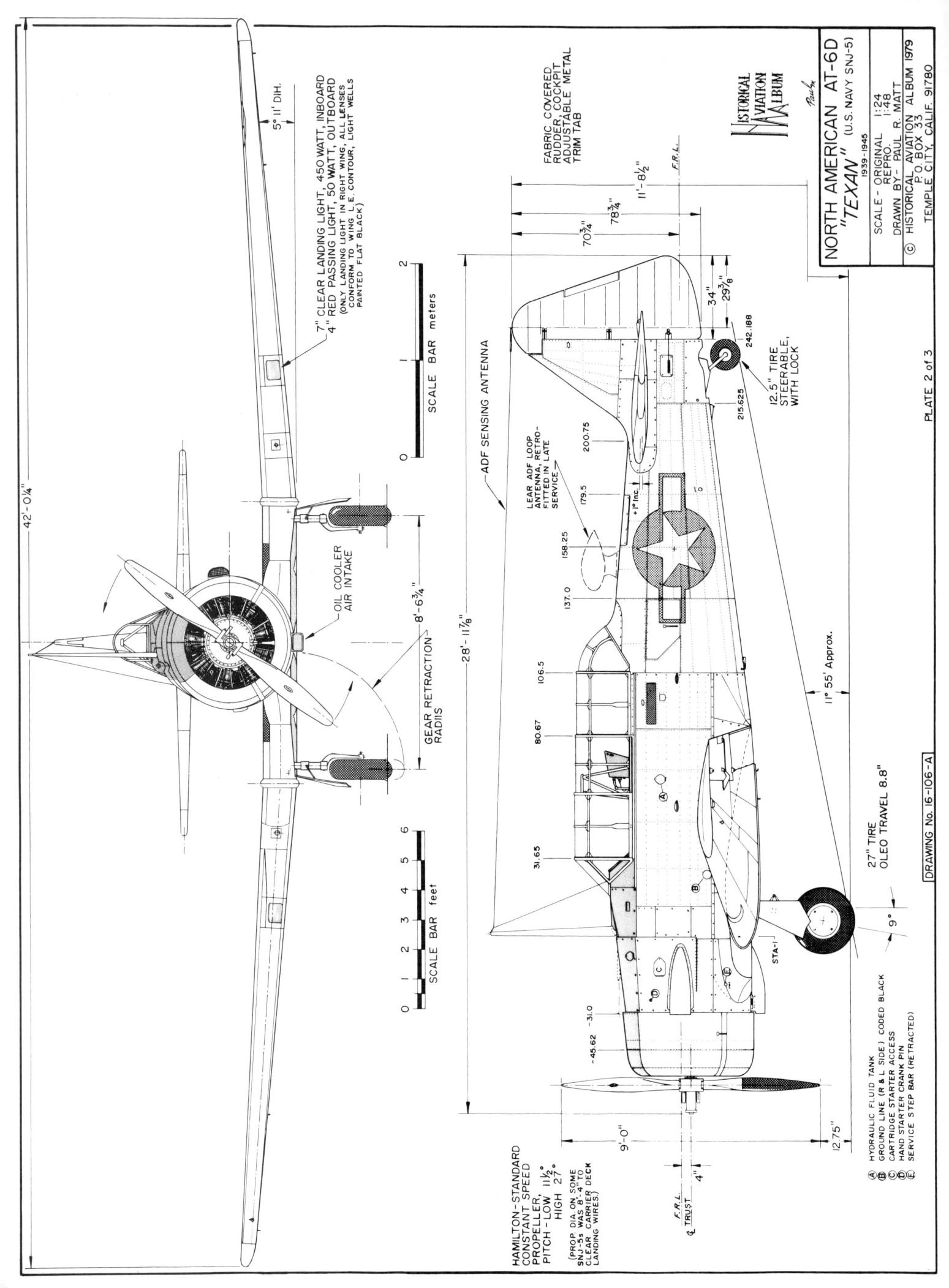

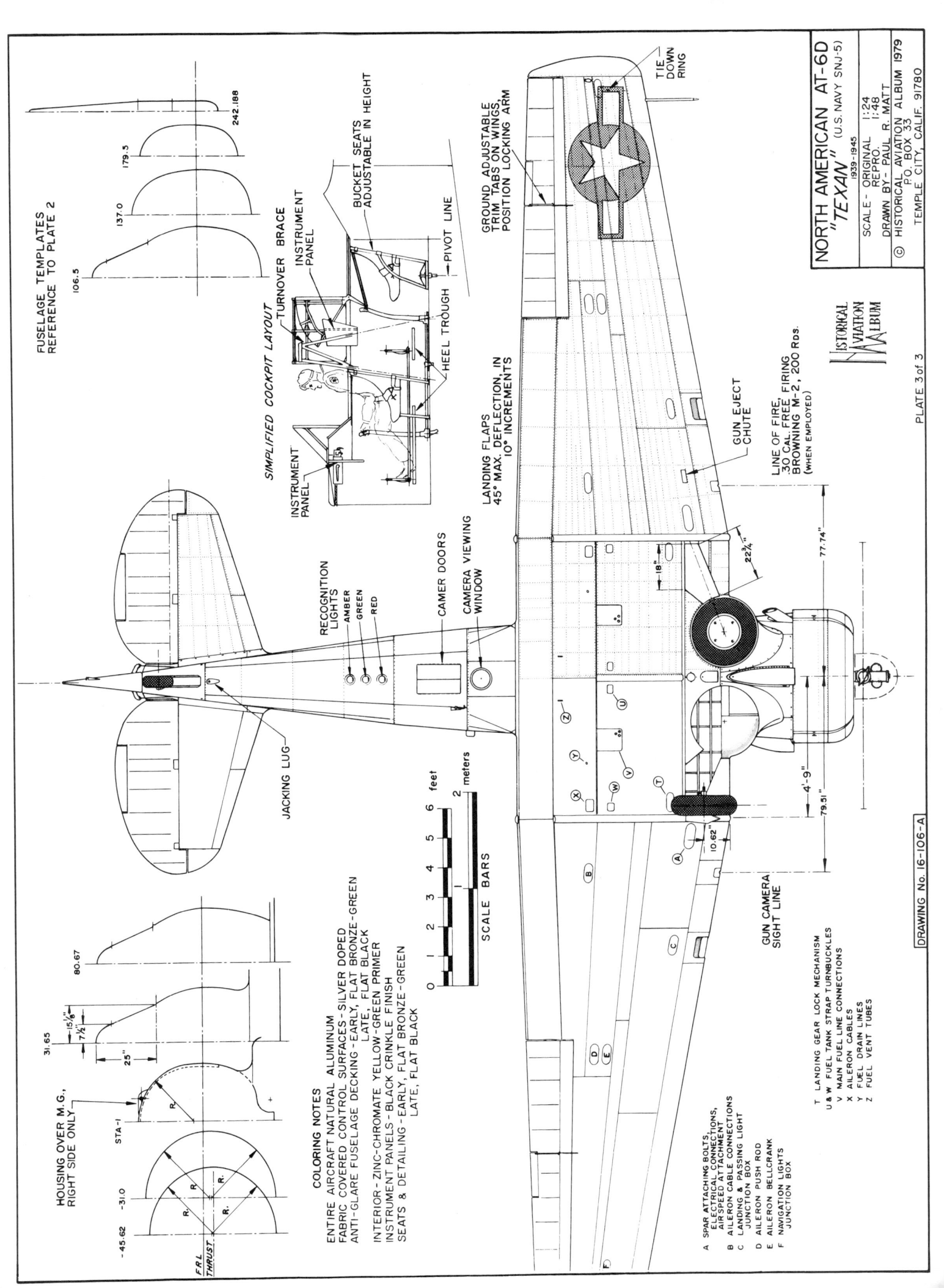

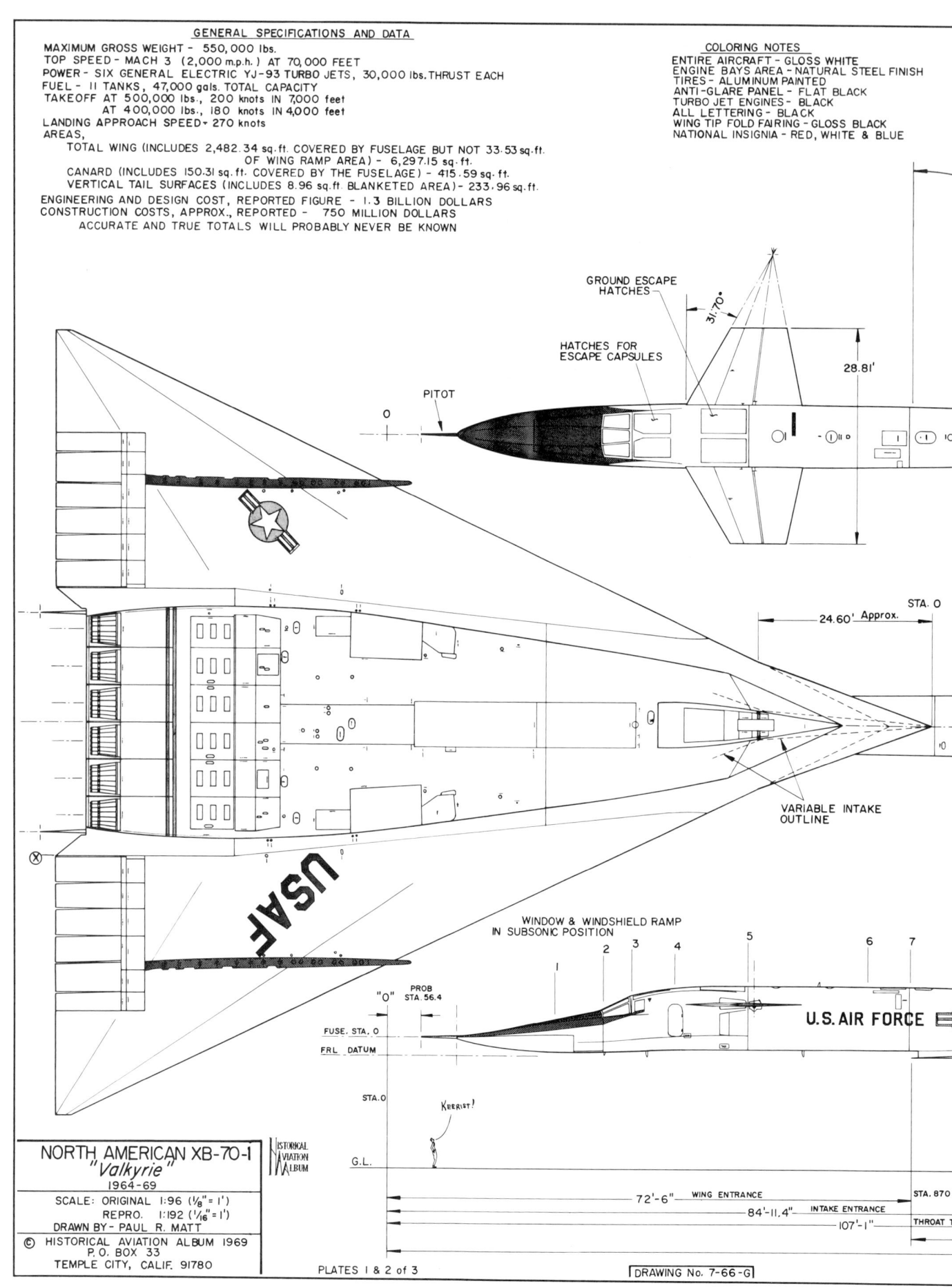

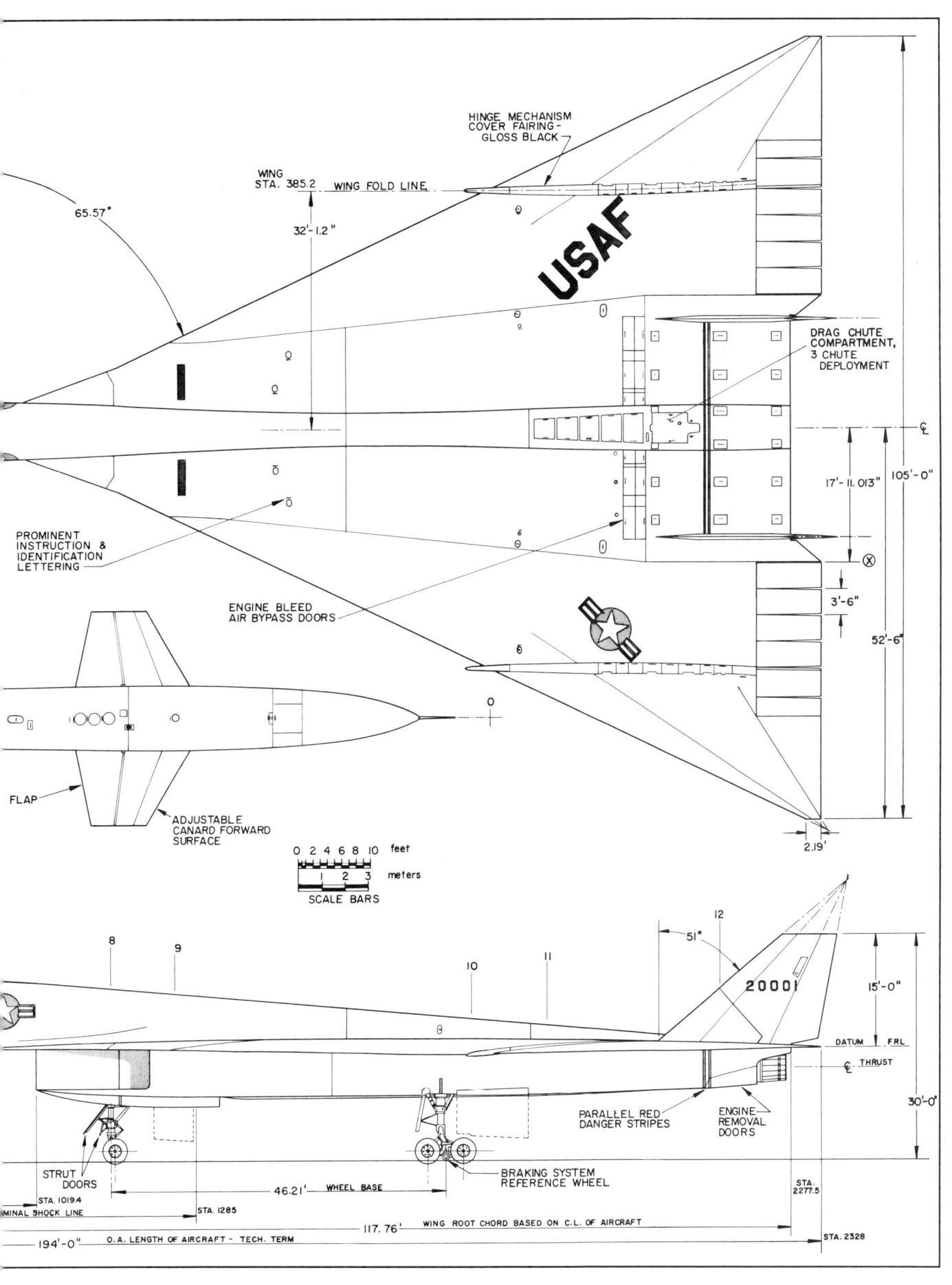

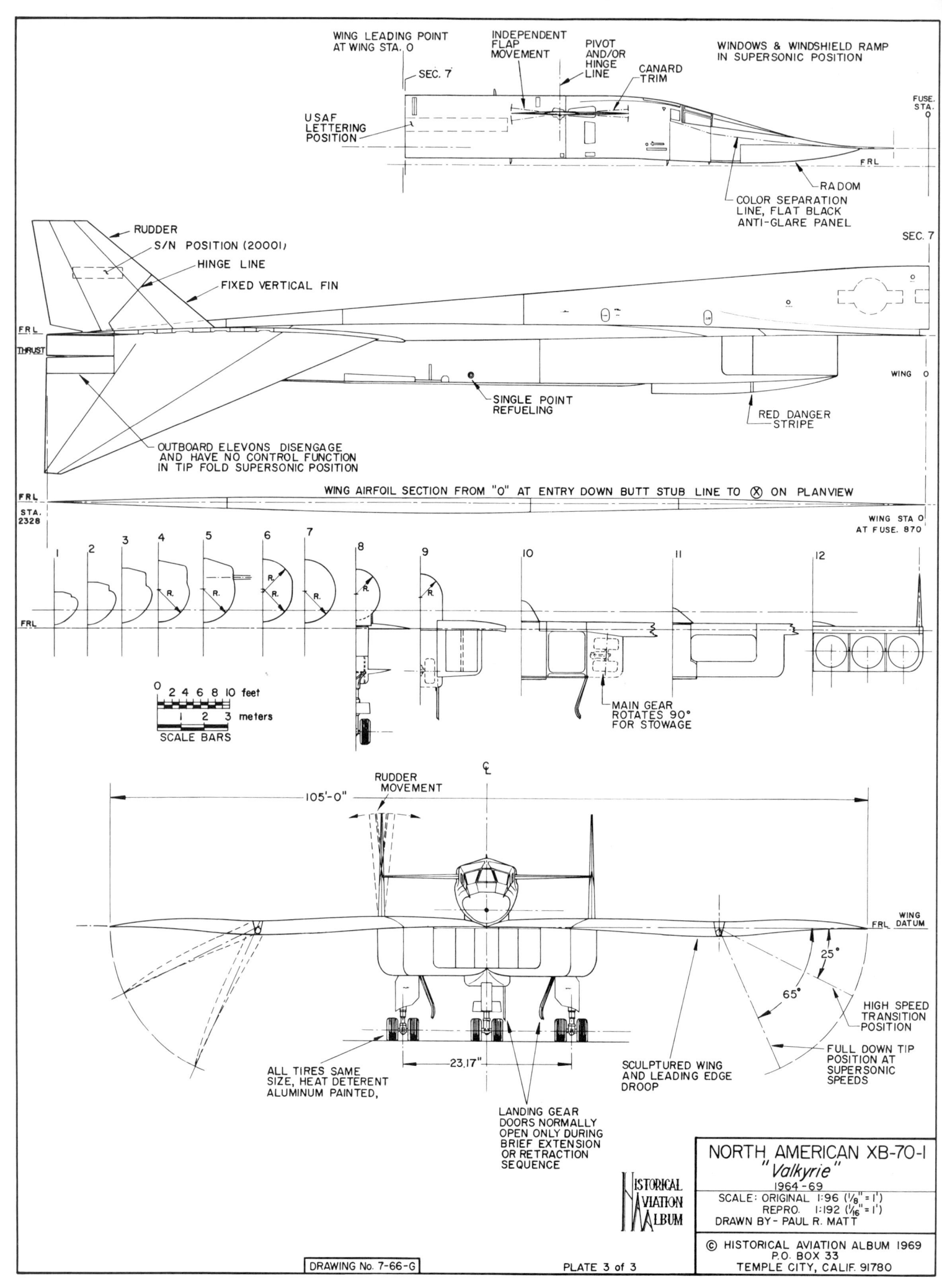

North American XB-70-1 "Valkyrie"

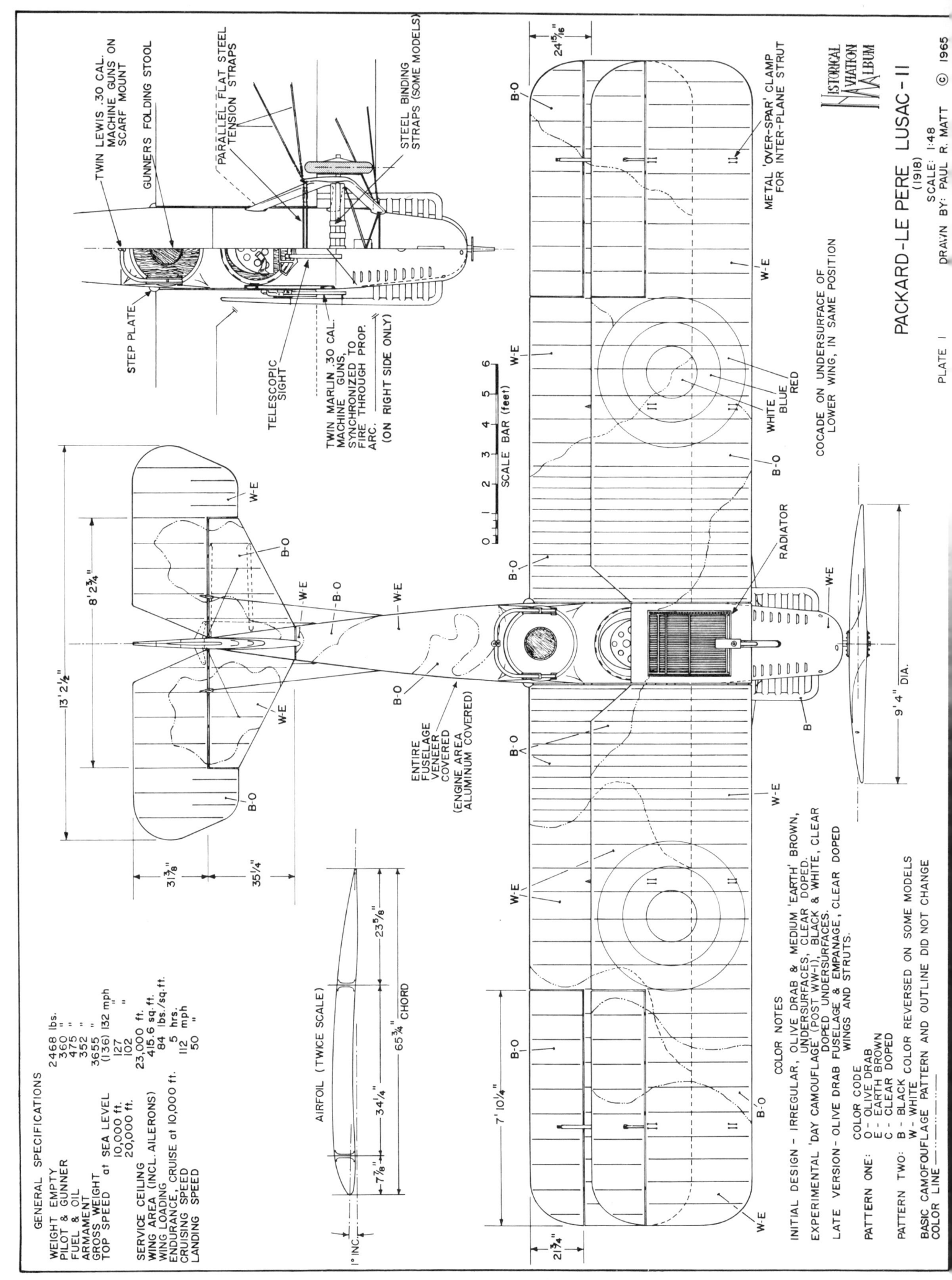

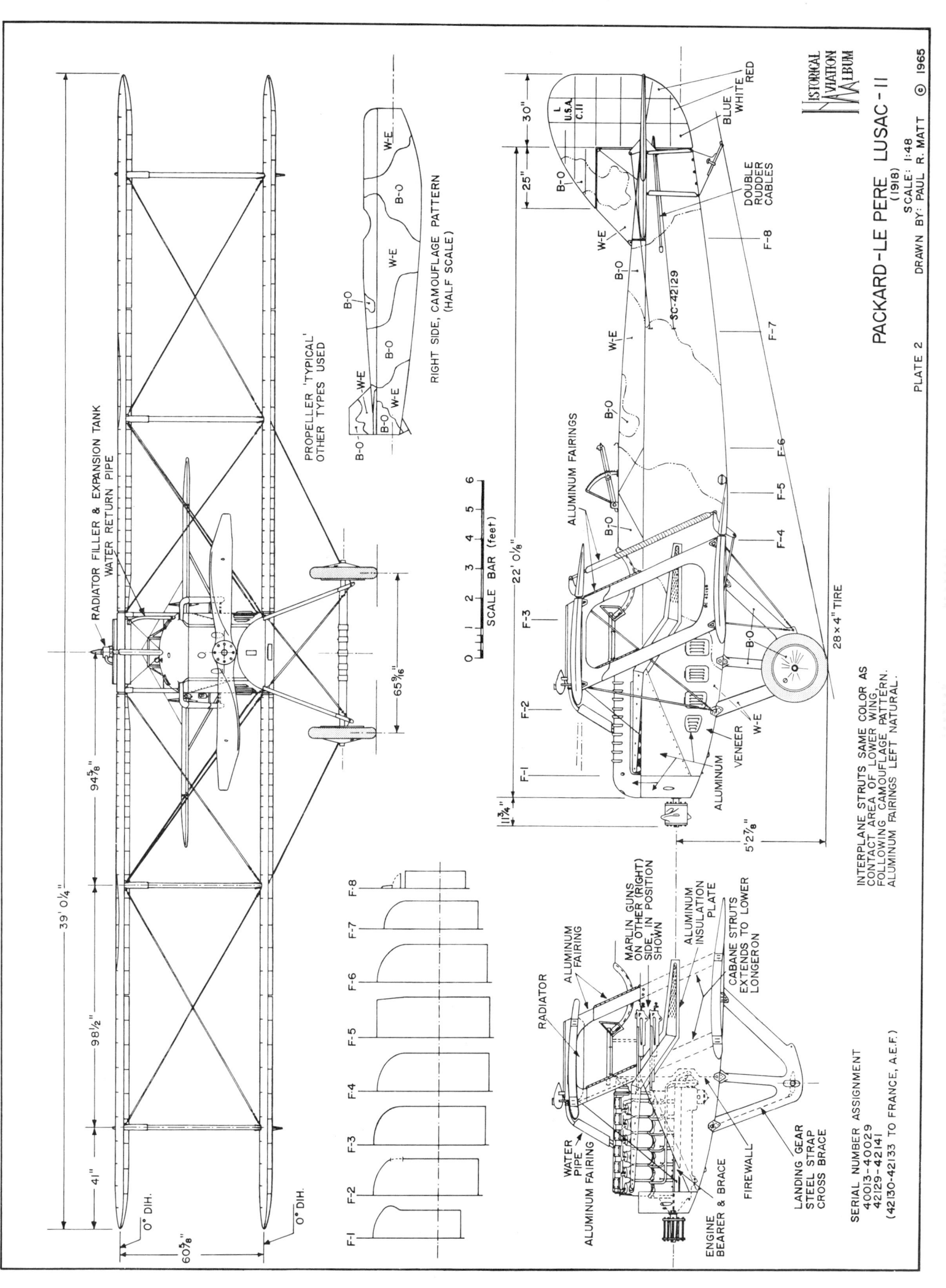

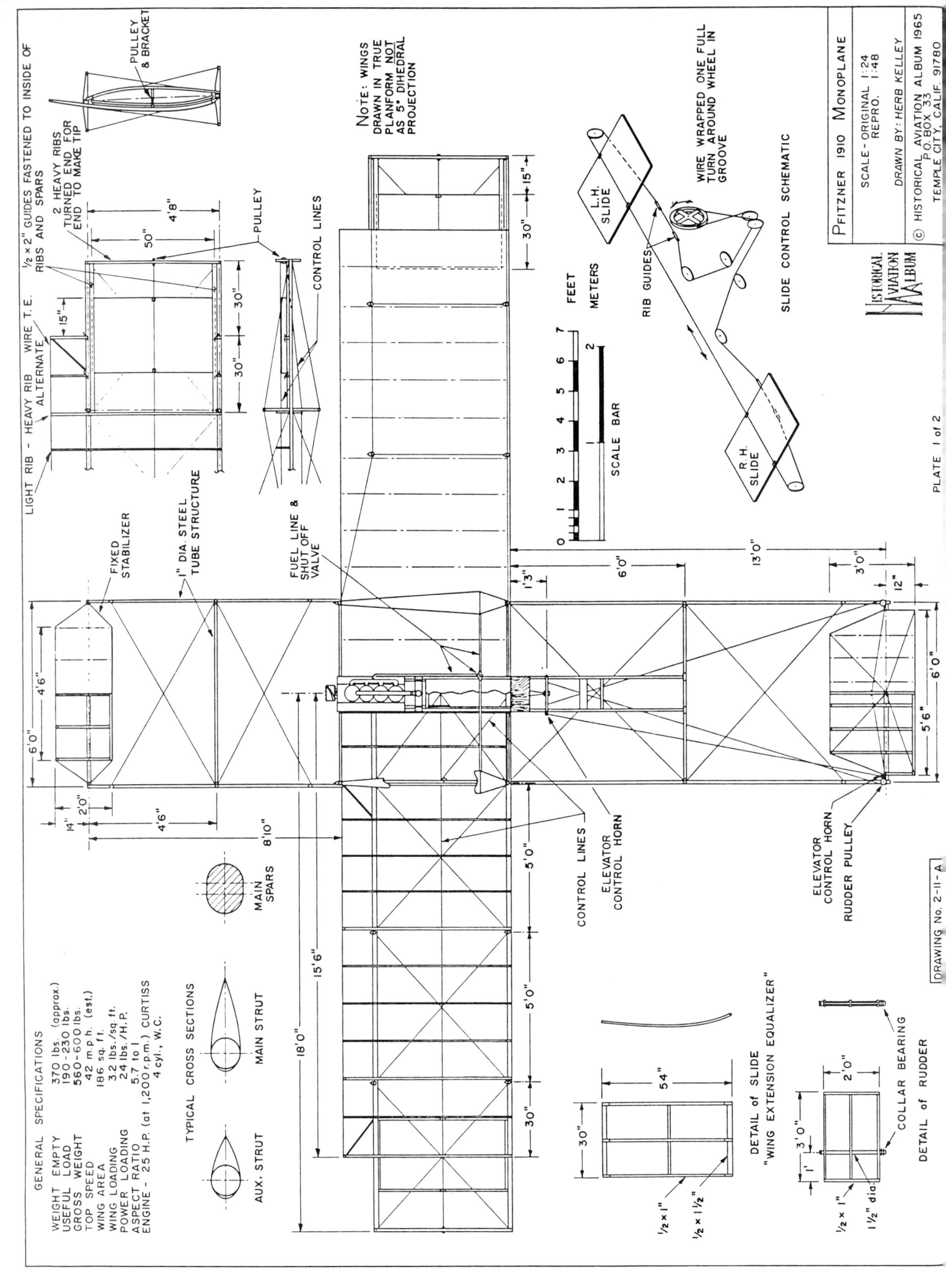

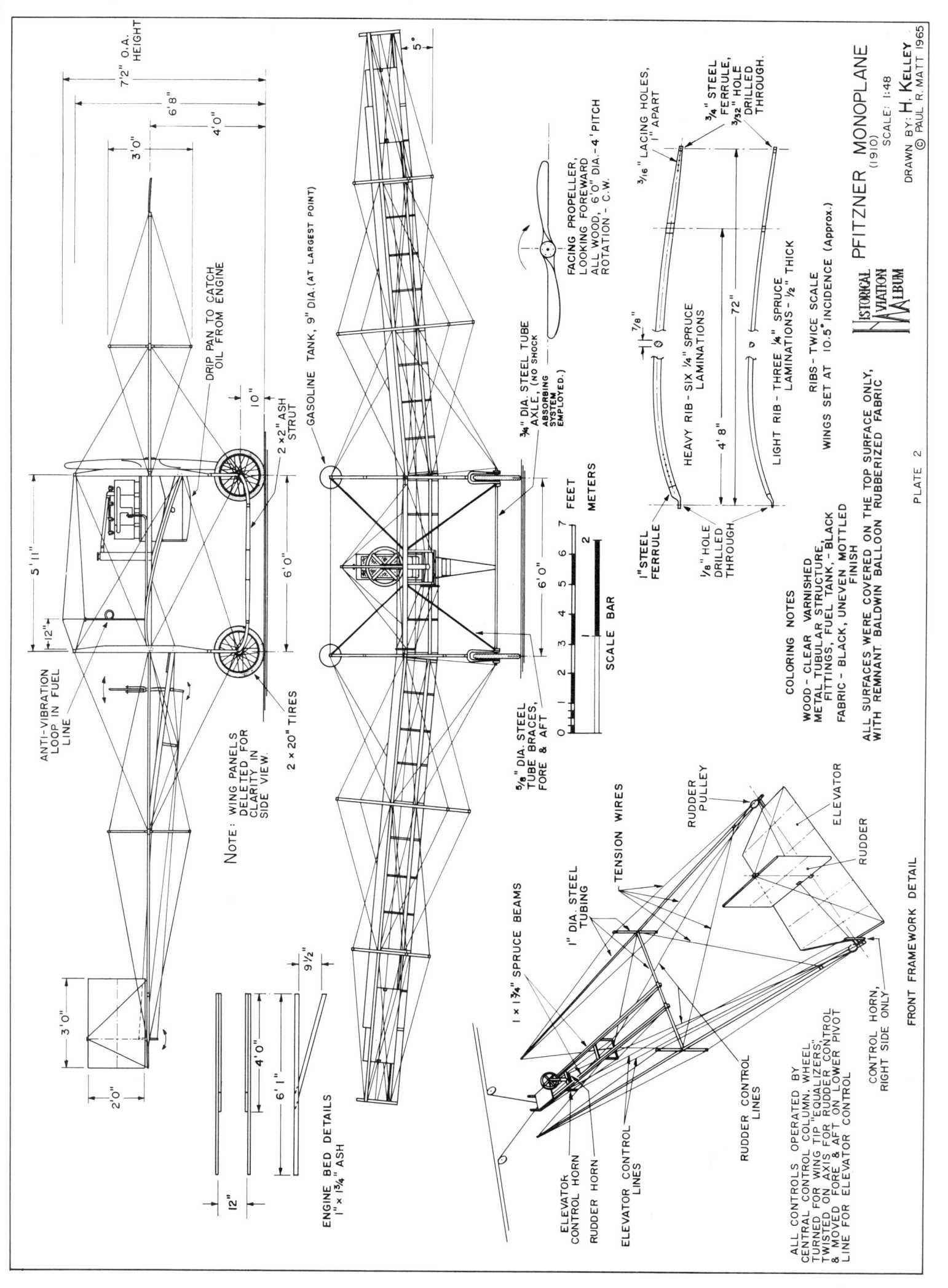

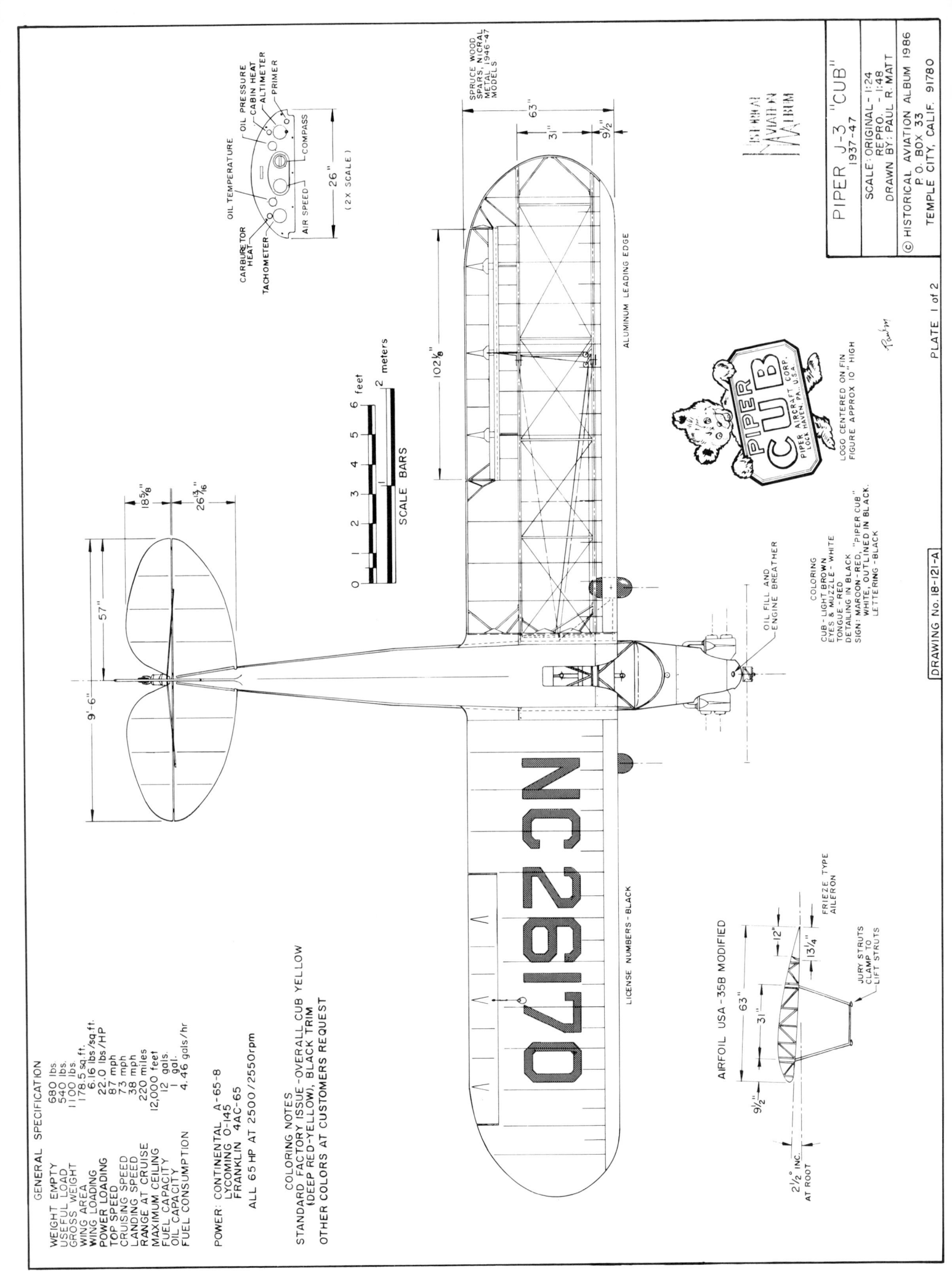

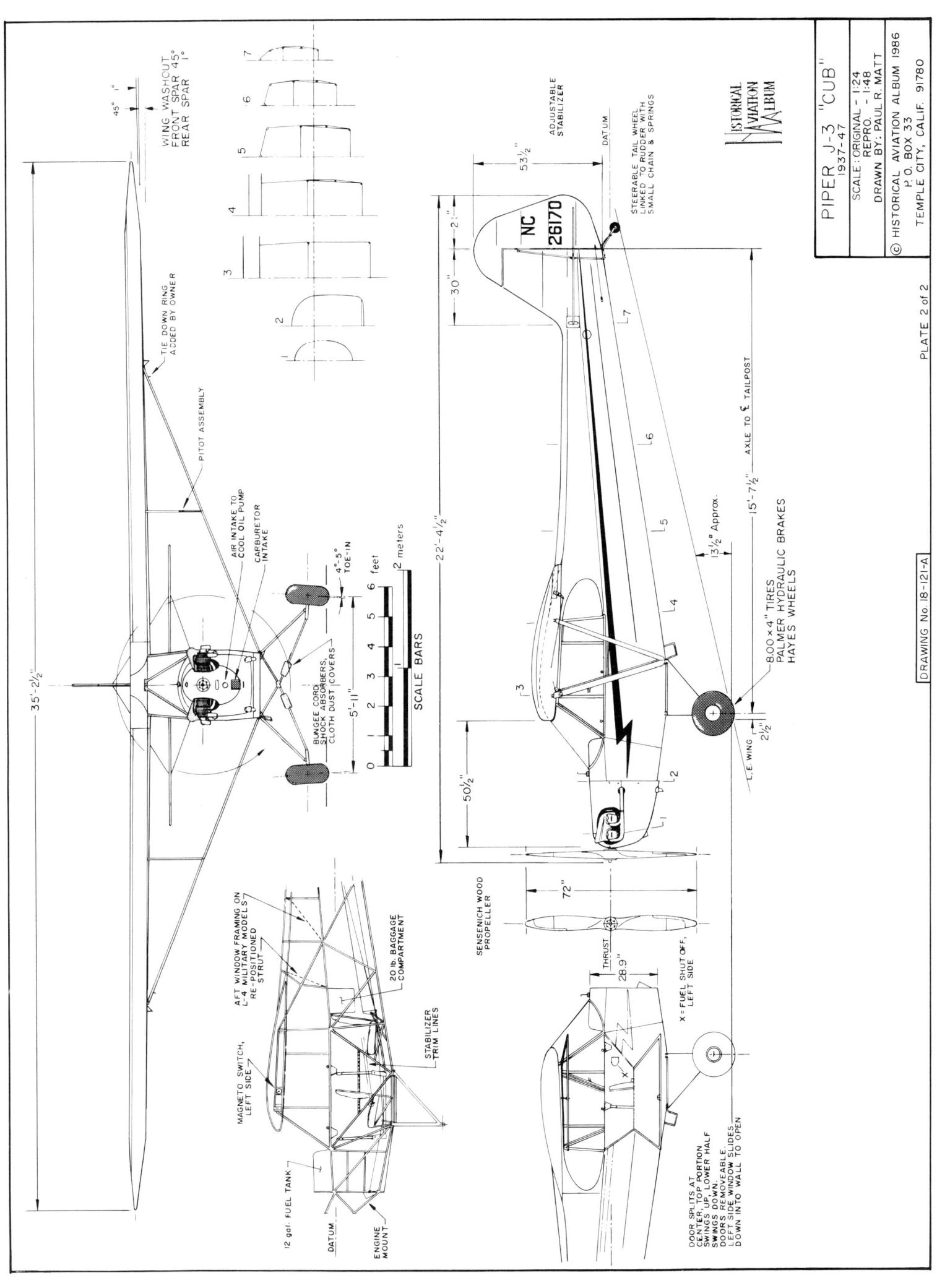

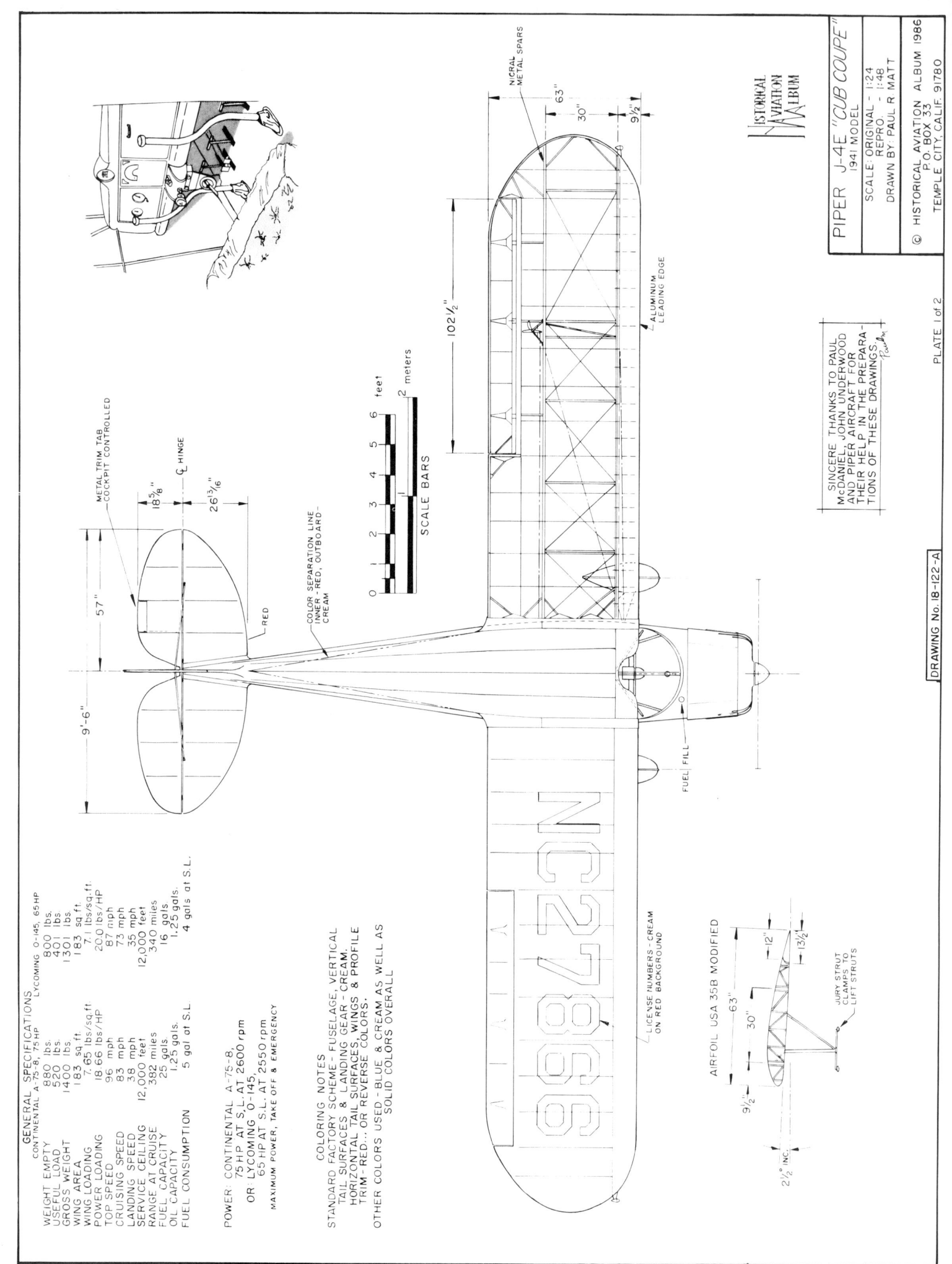

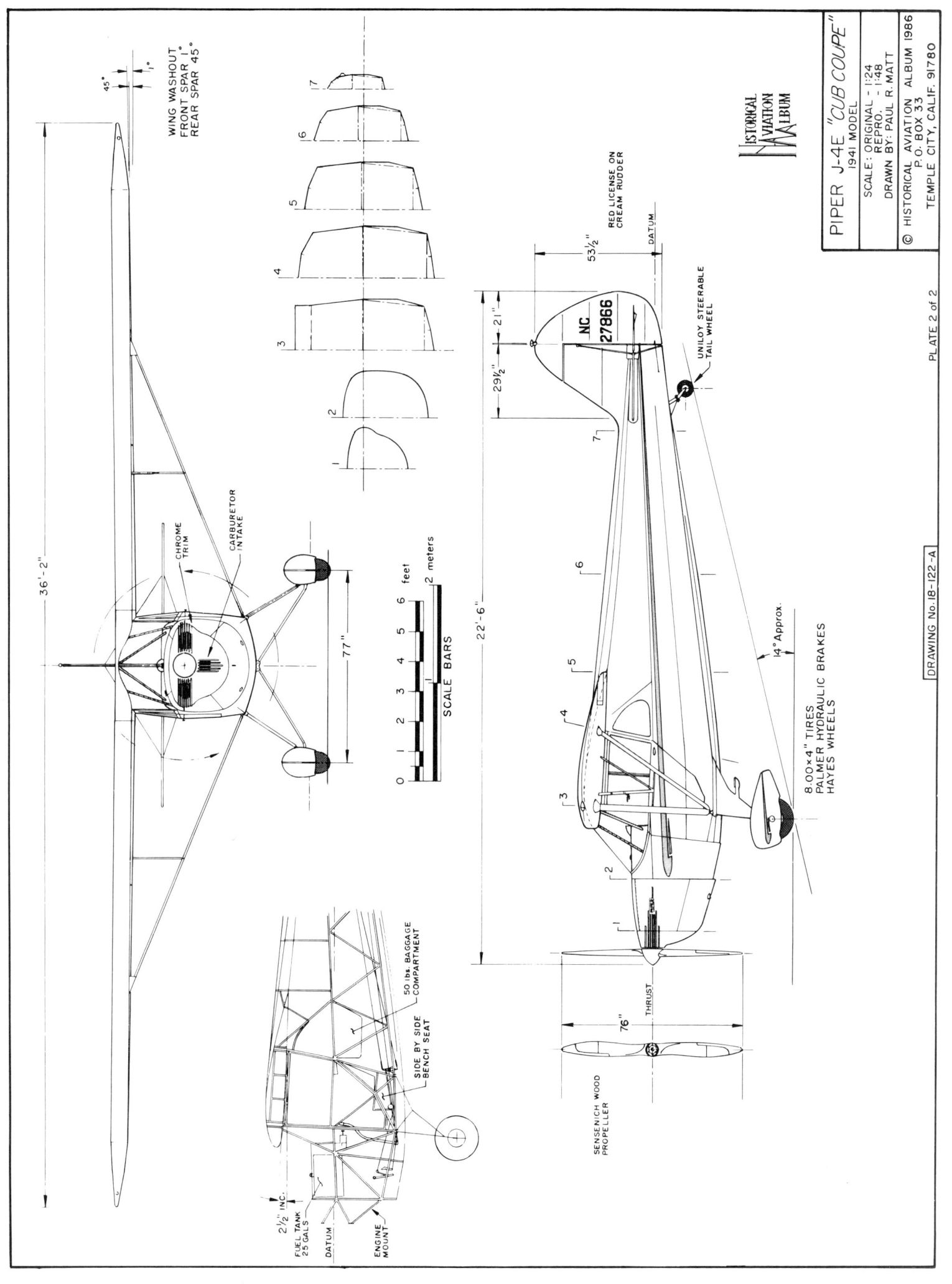

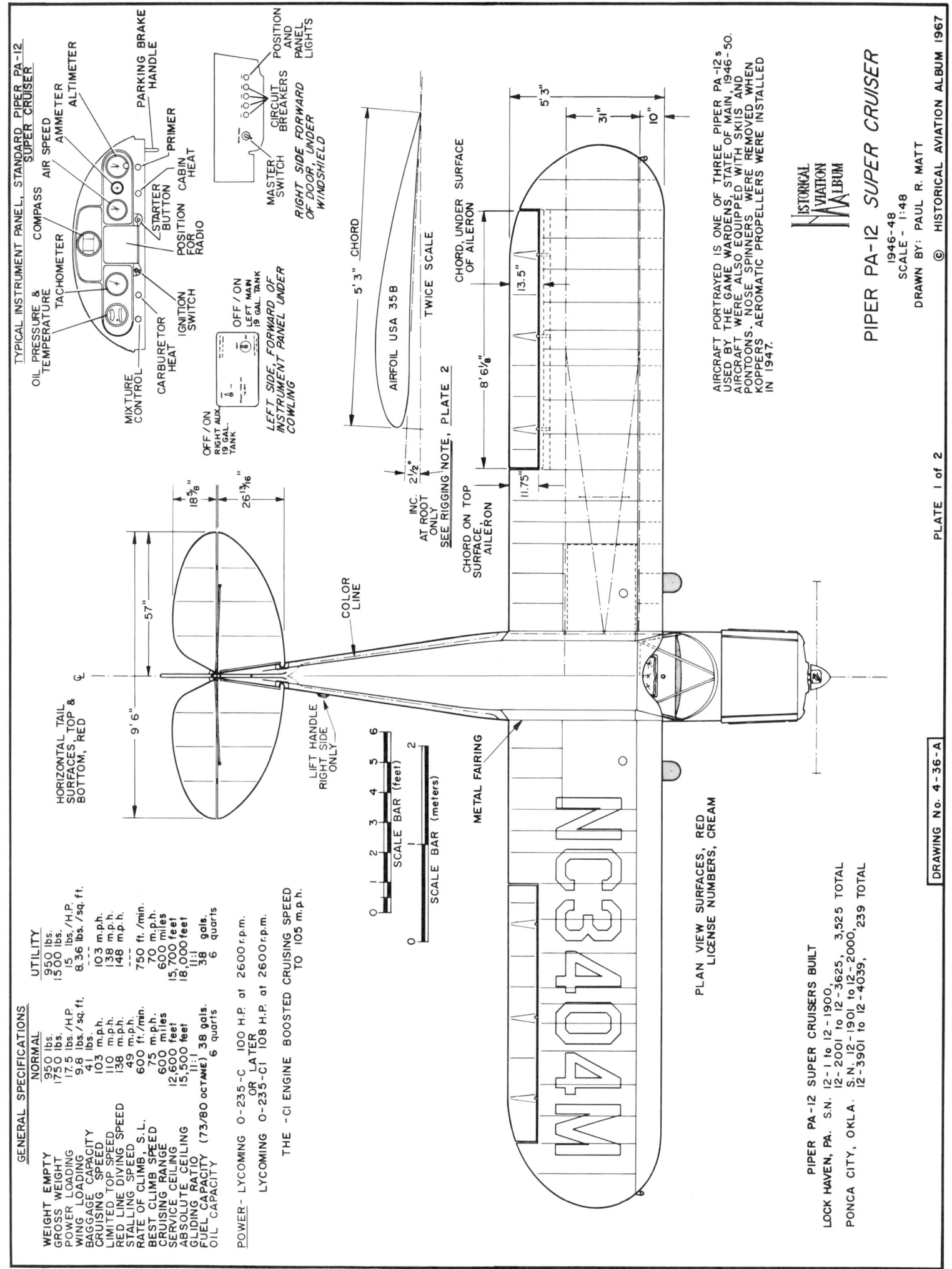

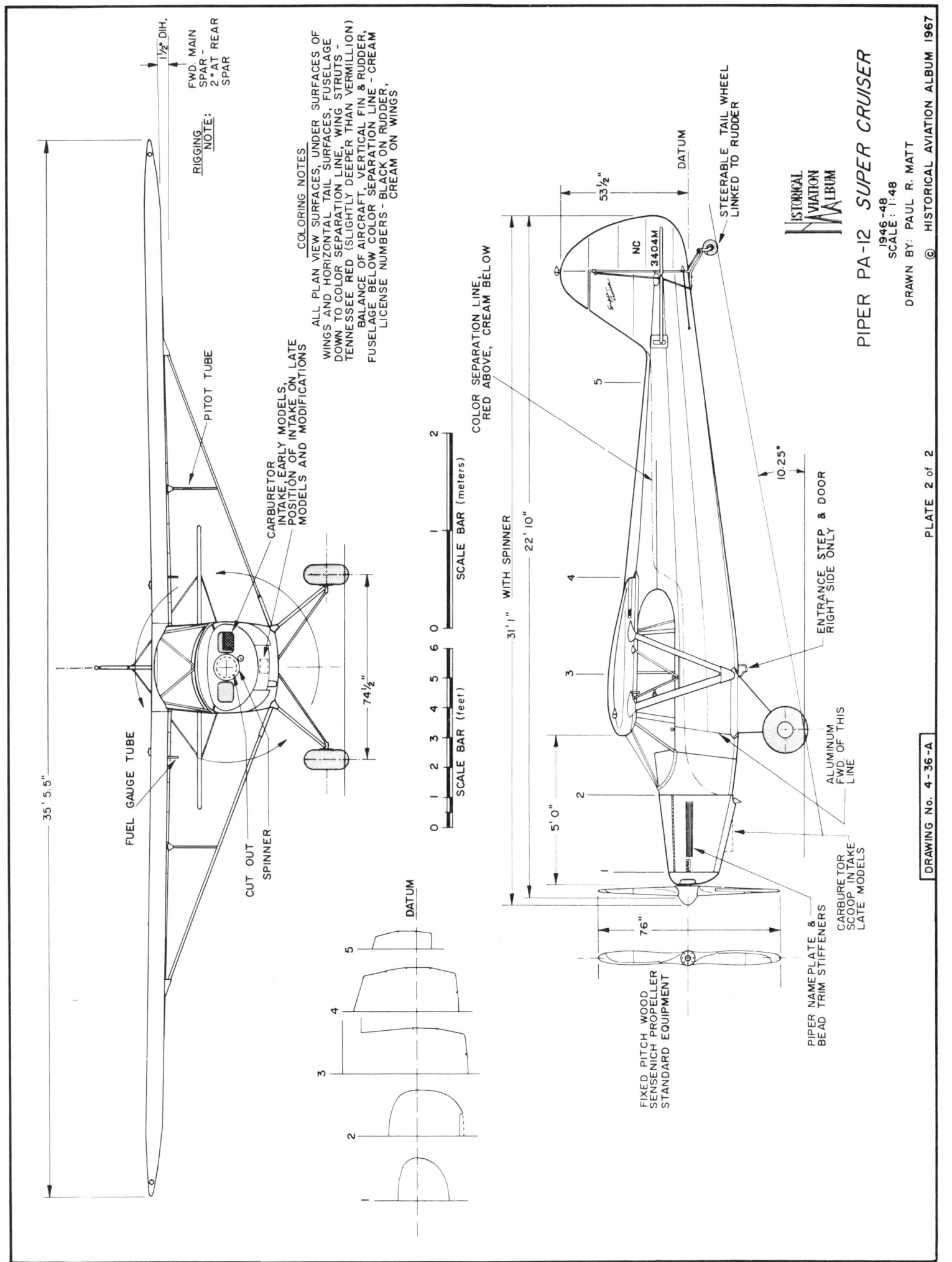

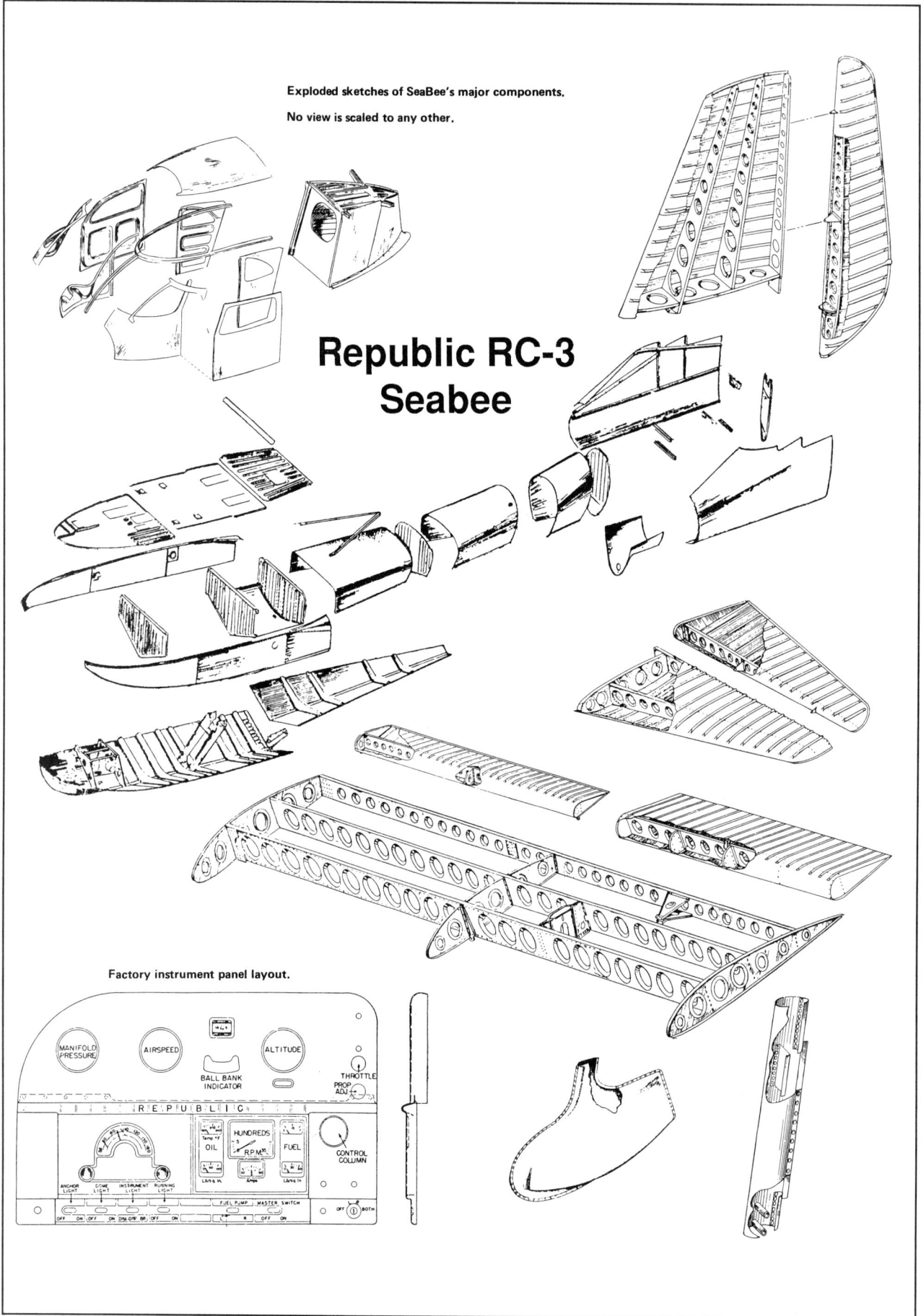

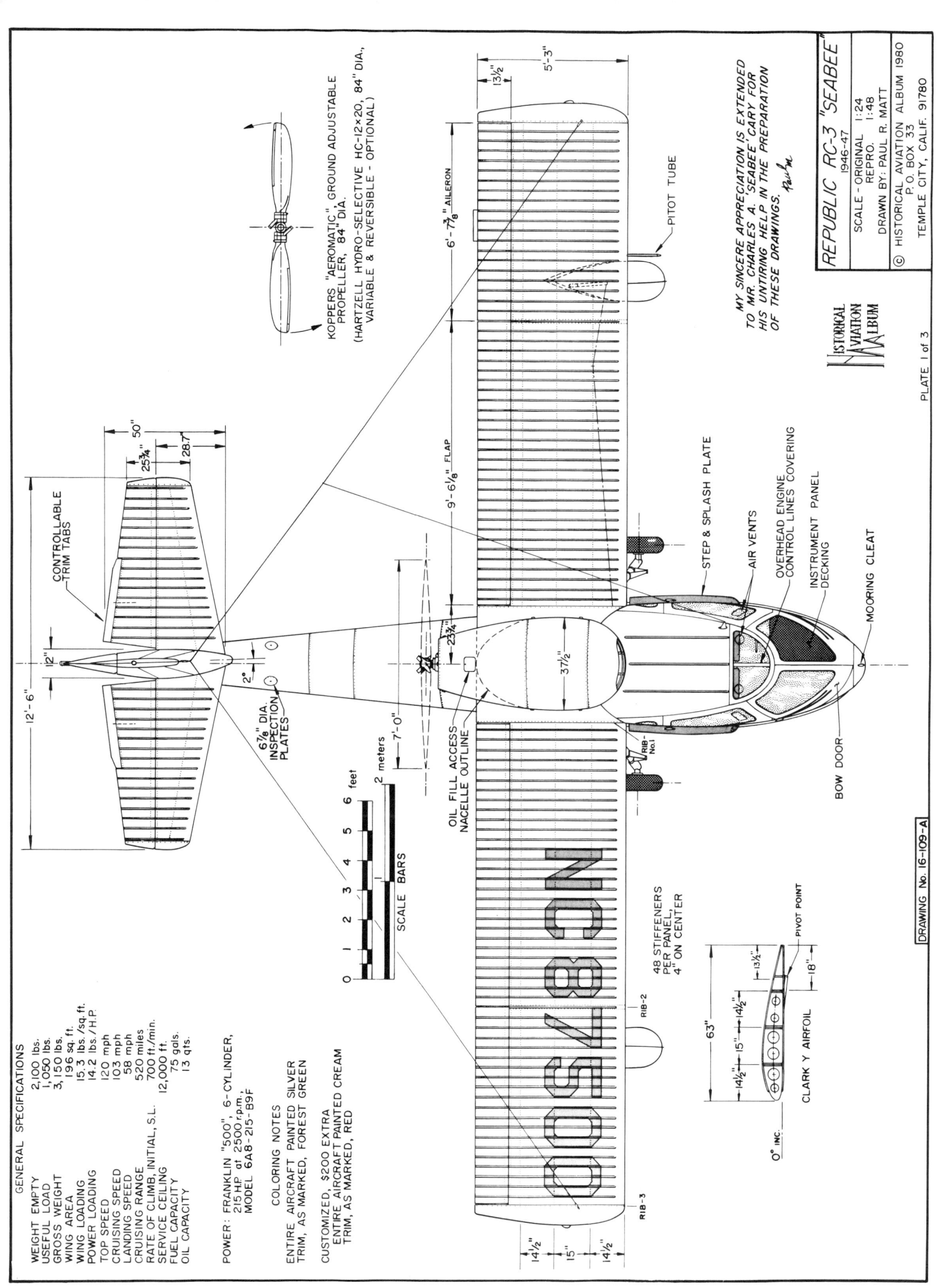

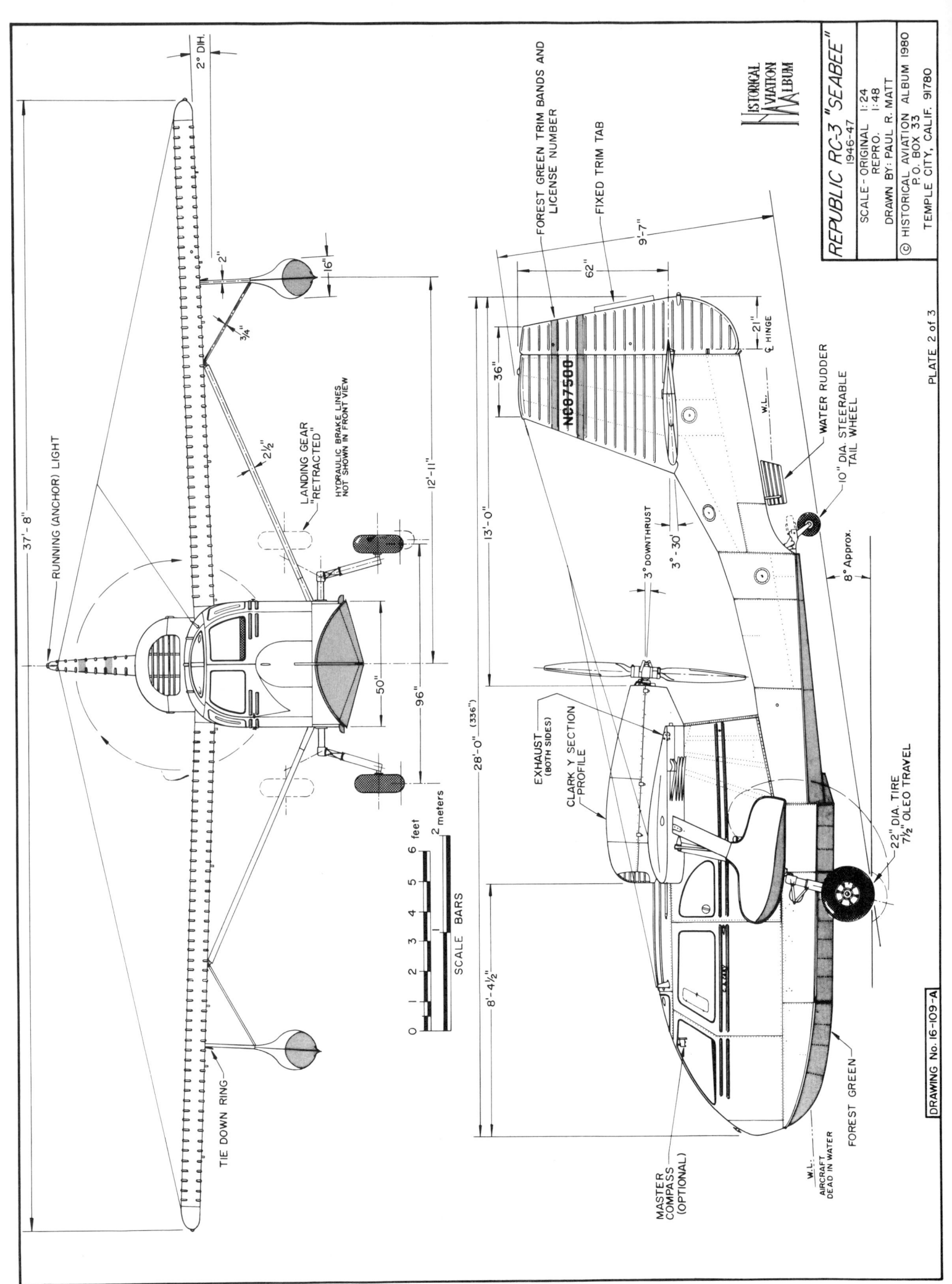

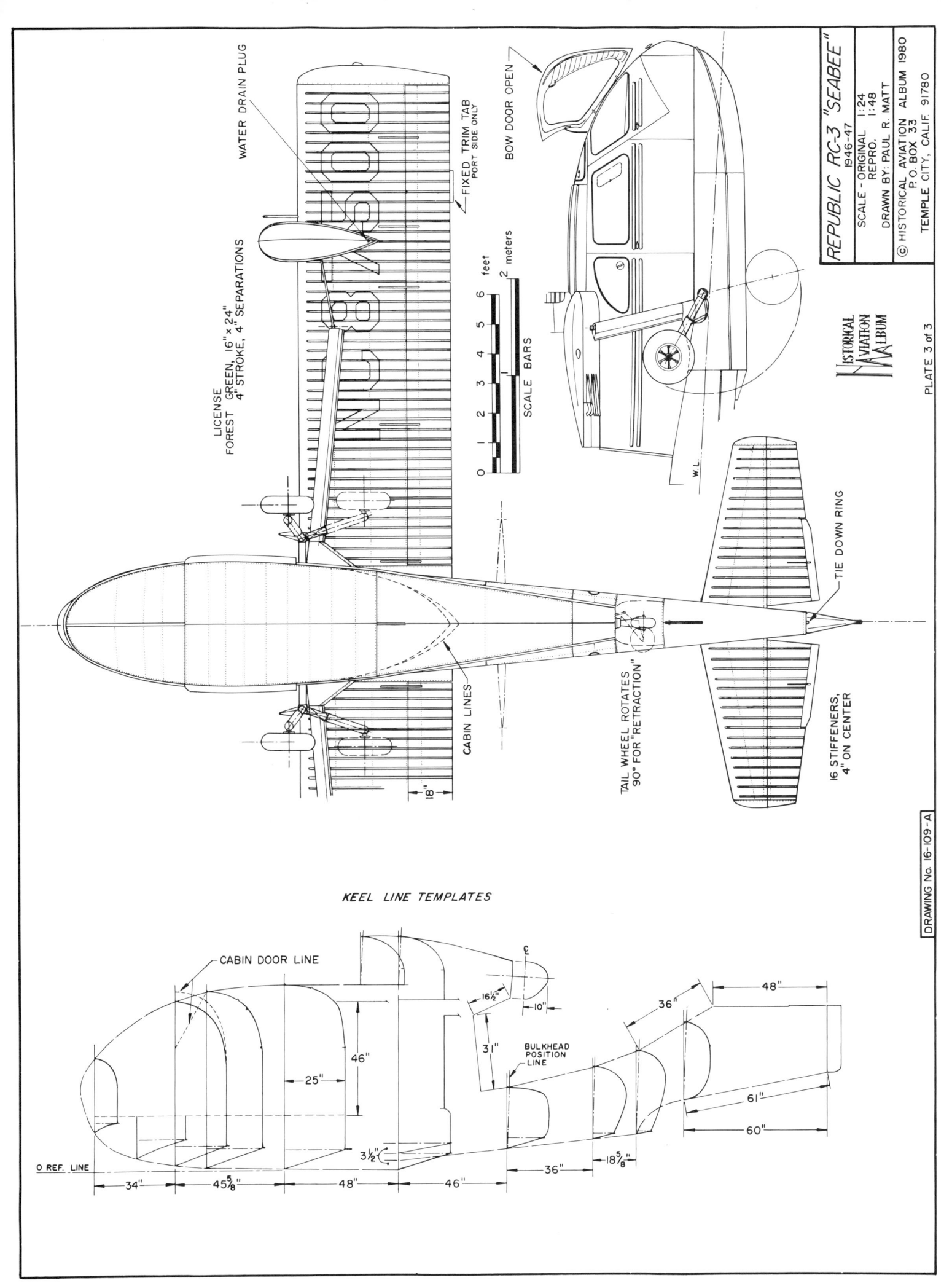

Rover Aviation Engine

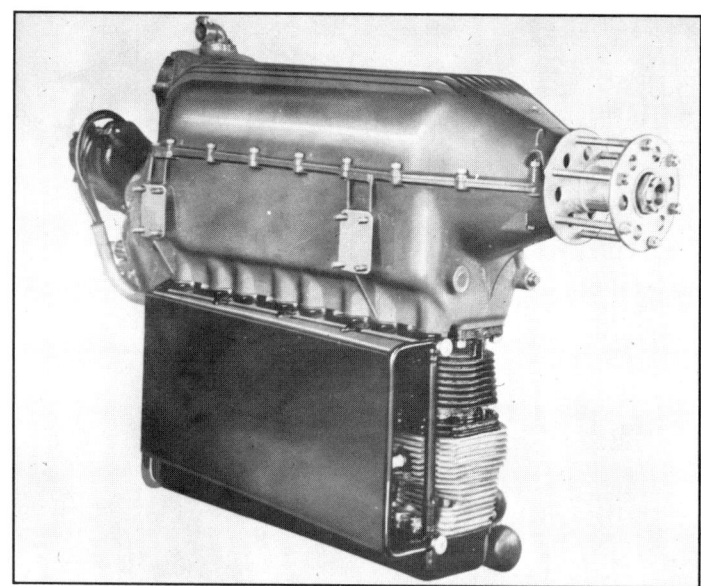

Rover Aviation Engine

Rover engine installation in Driggs "Skylark".

Fairchild 22 with Rover engine.

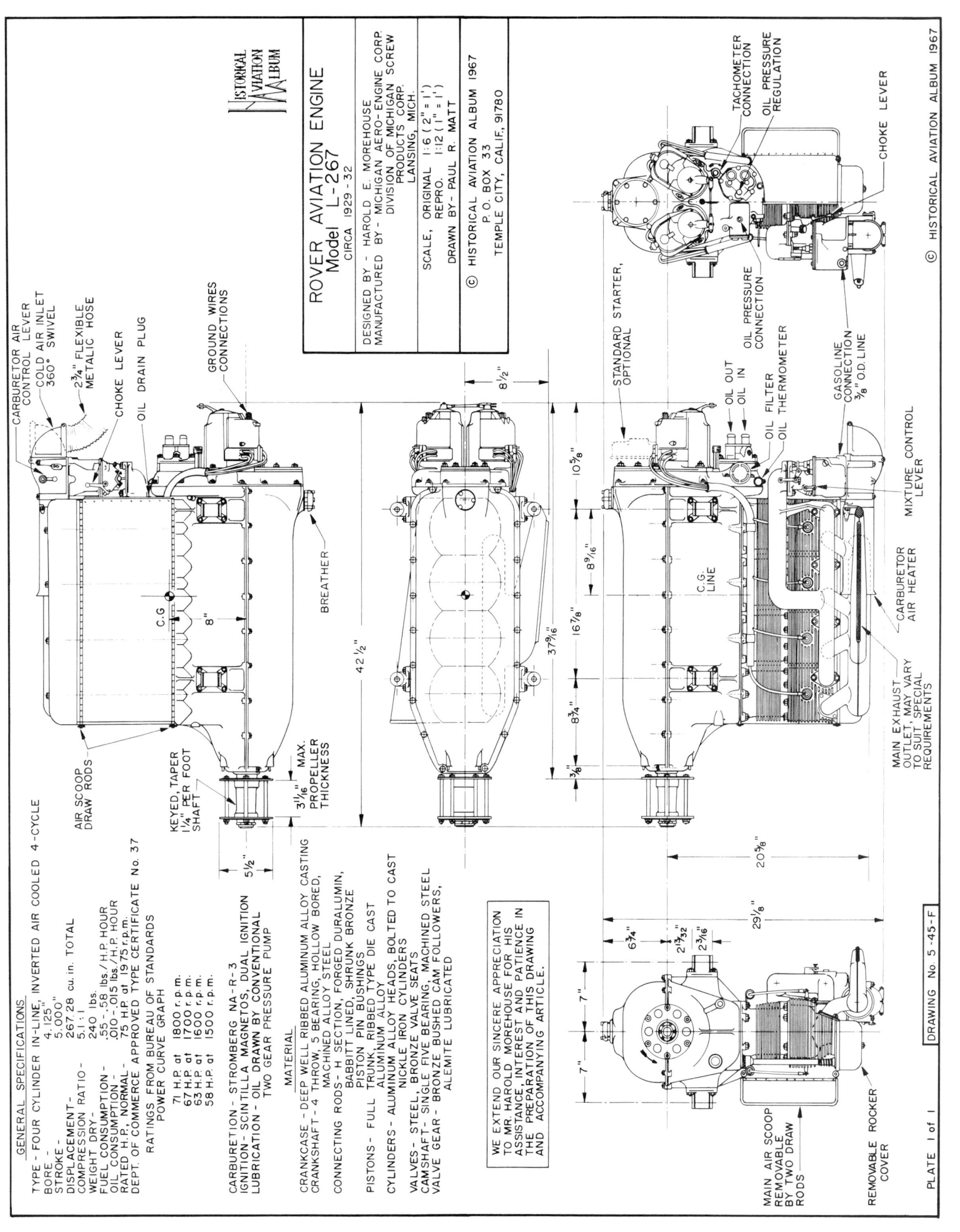

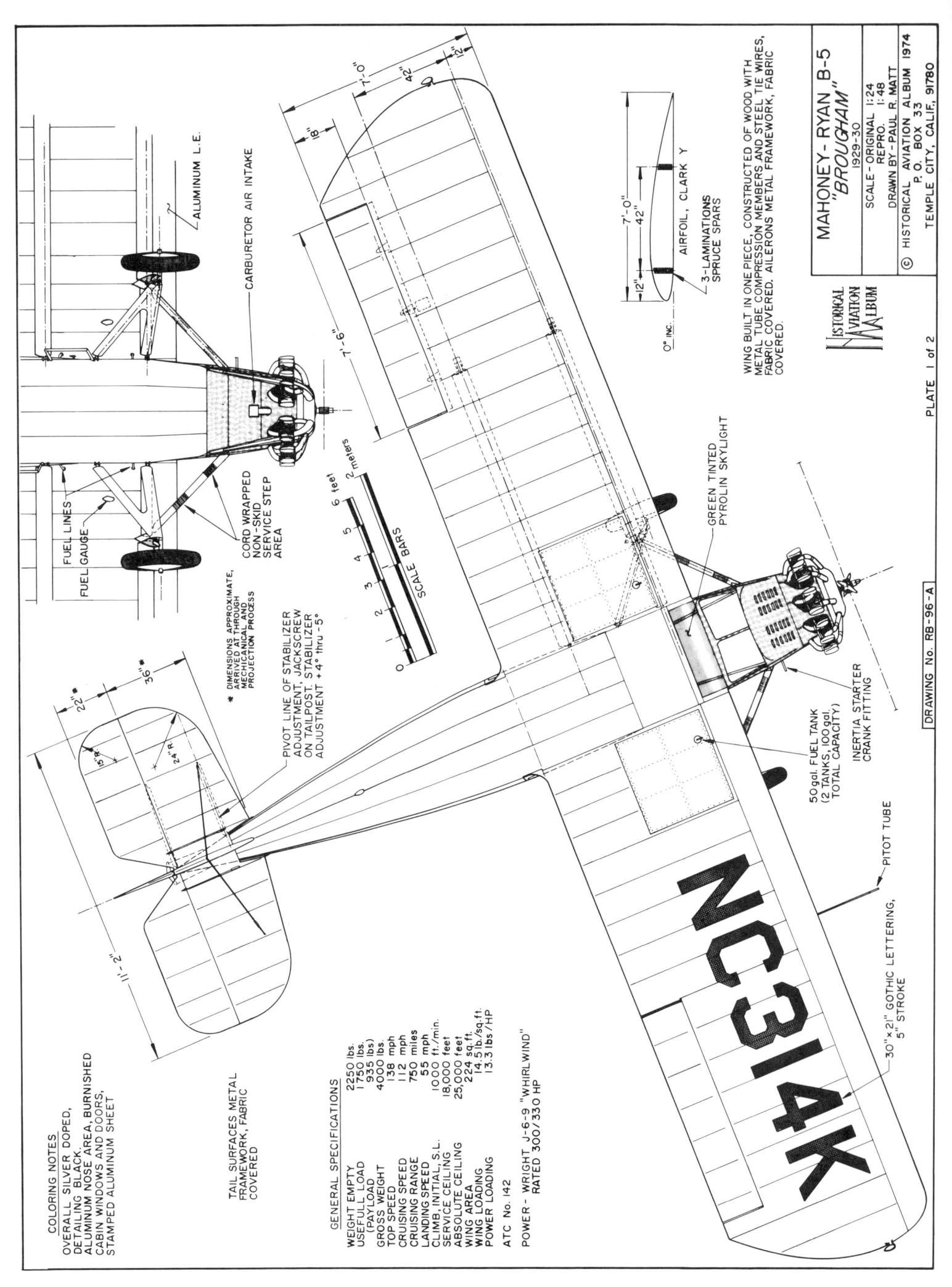

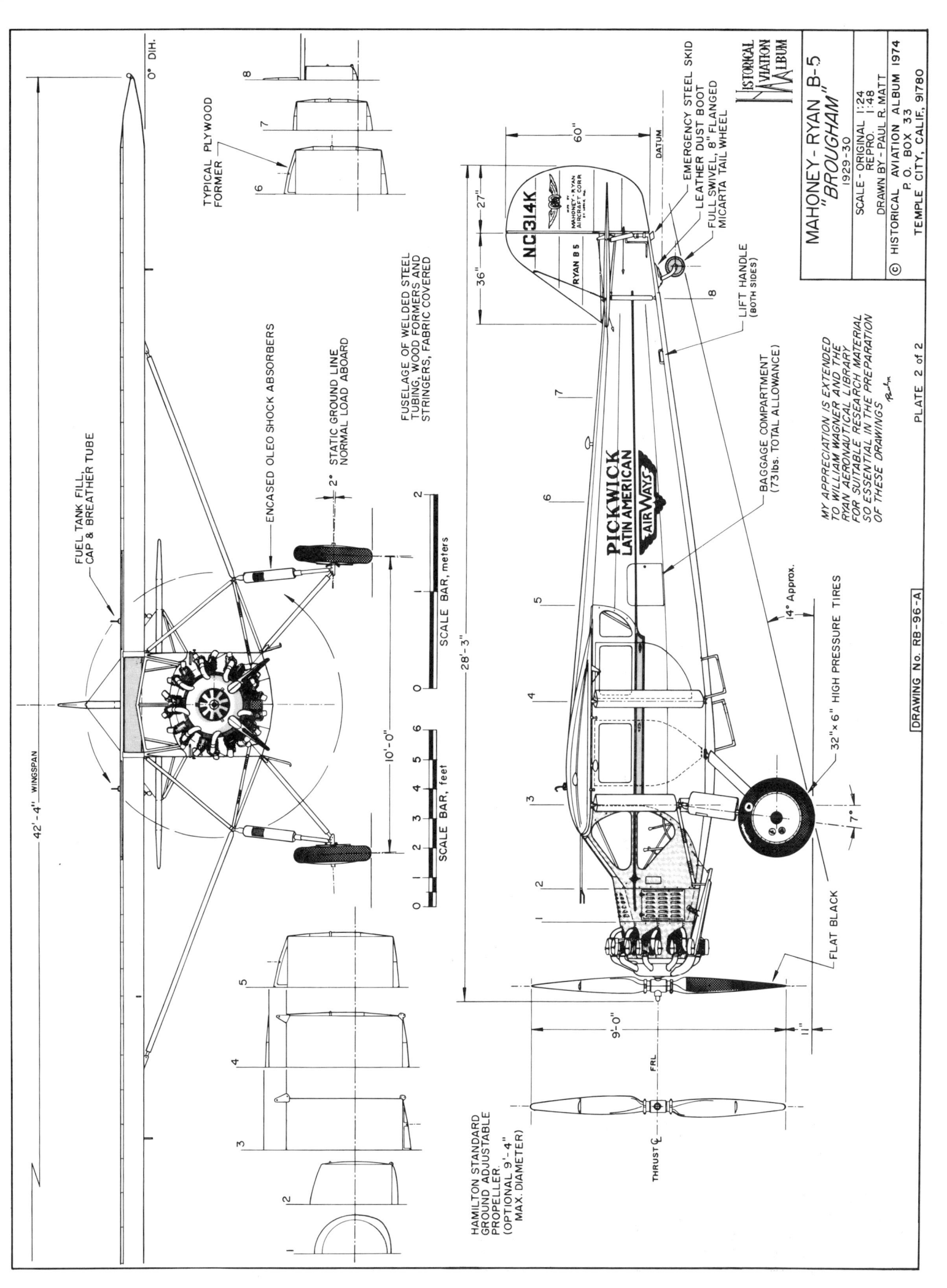

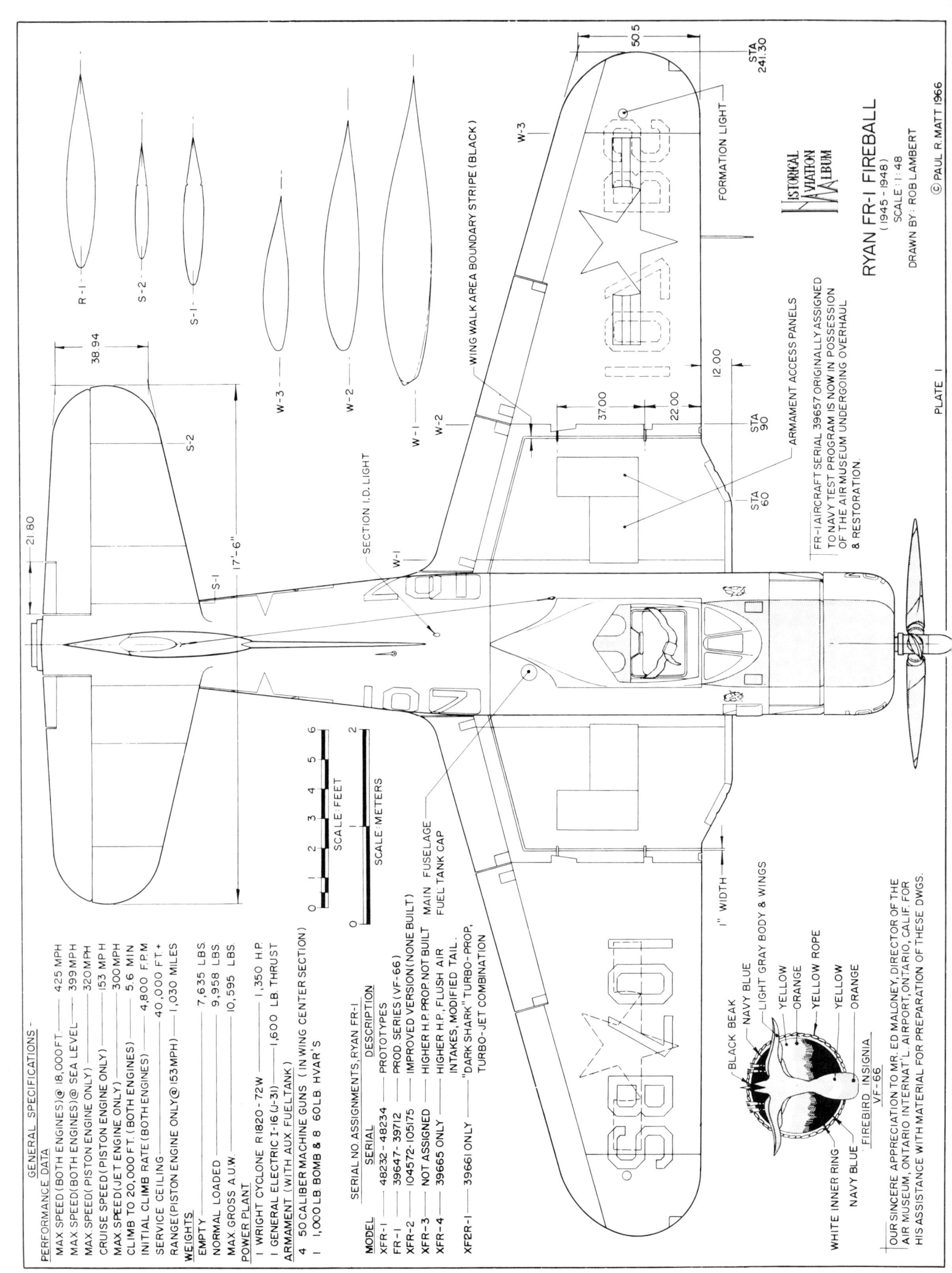

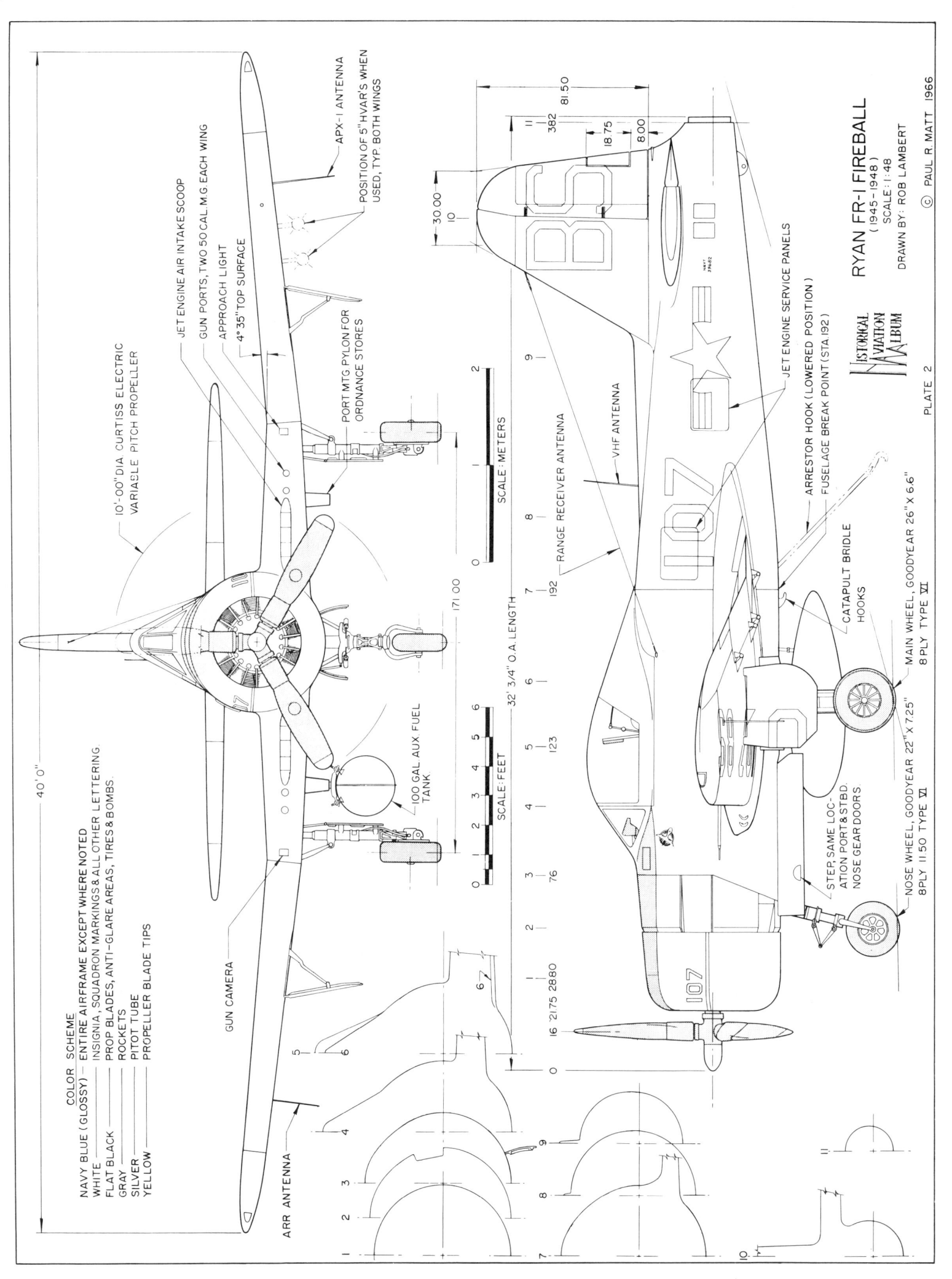

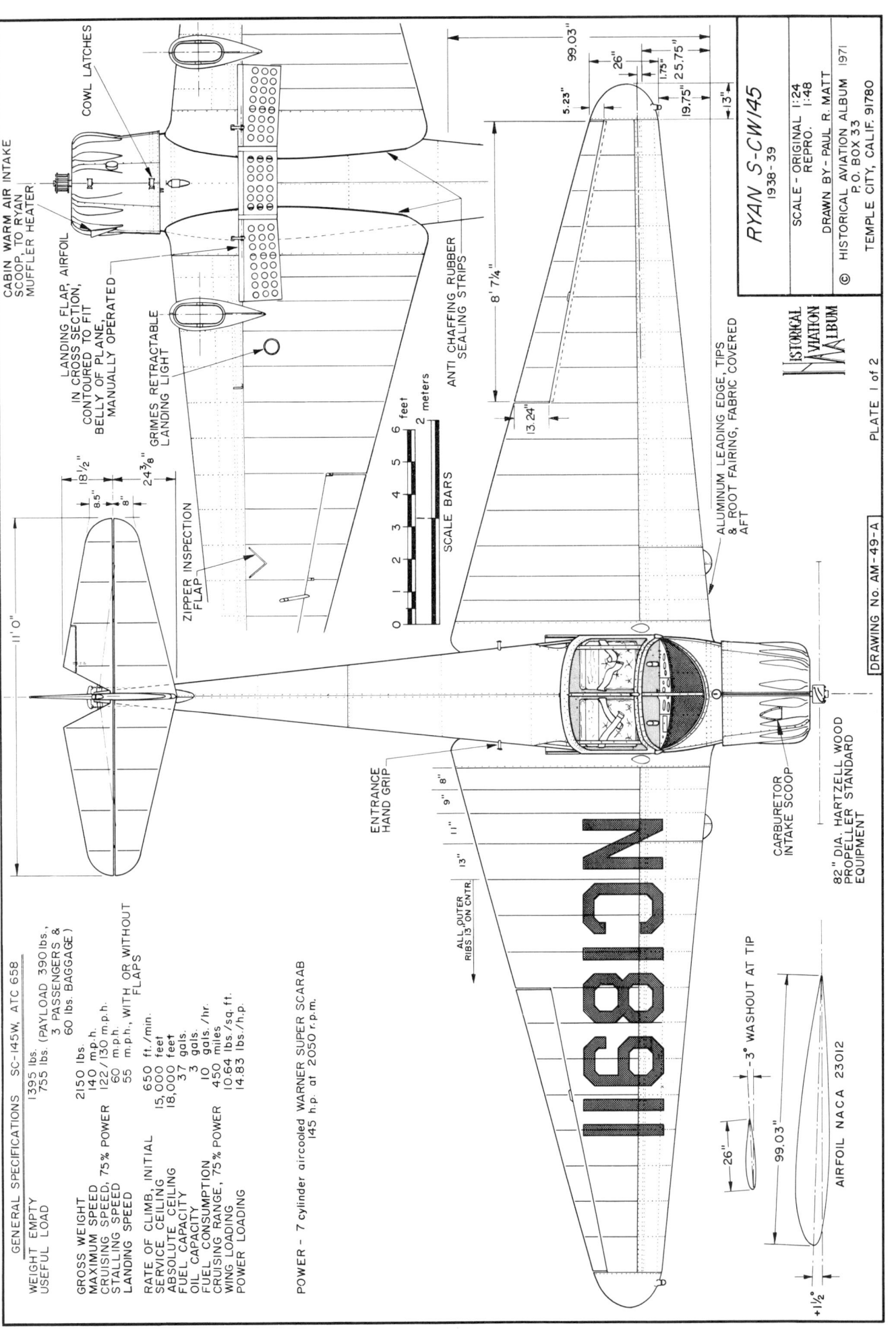

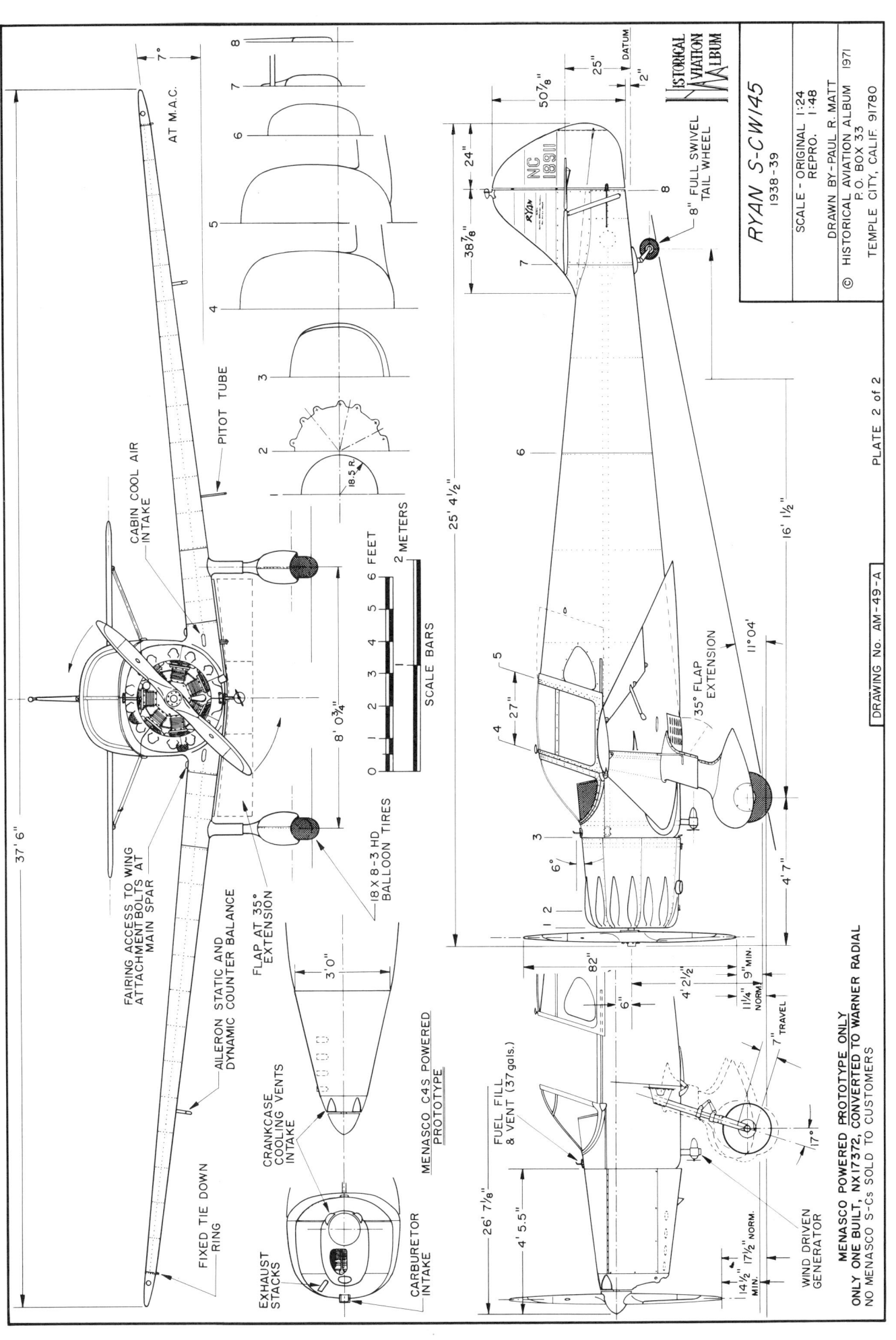

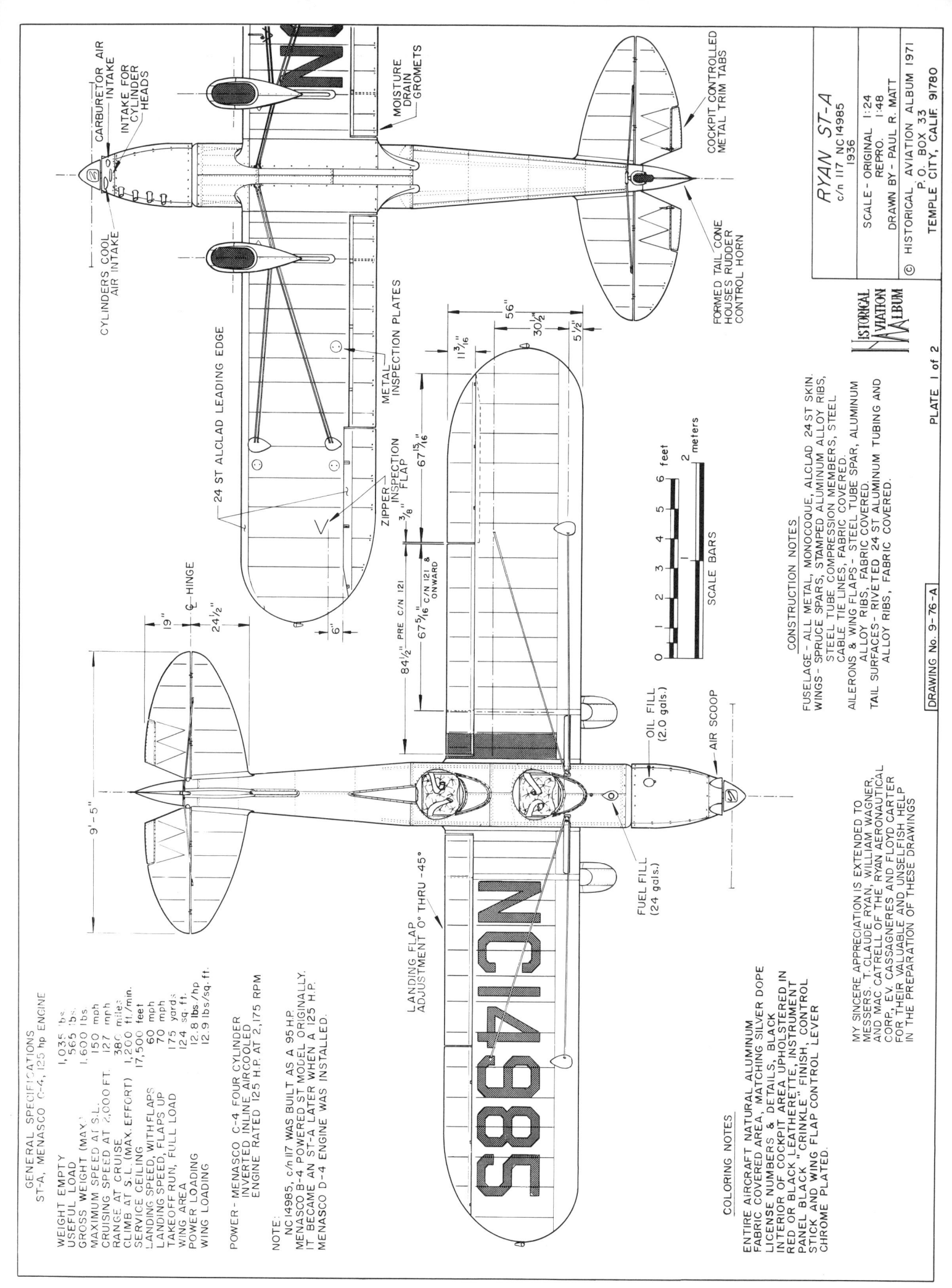

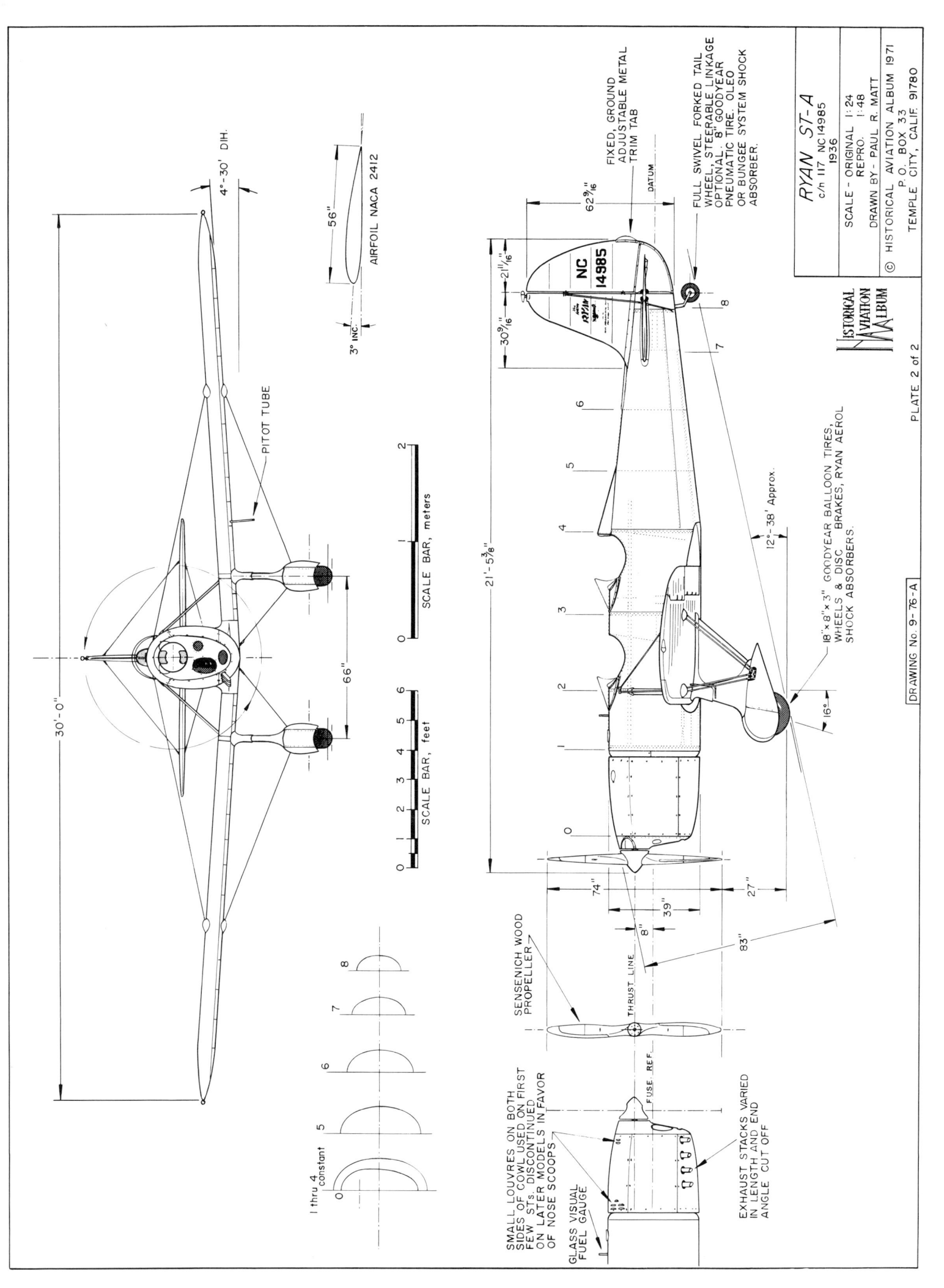

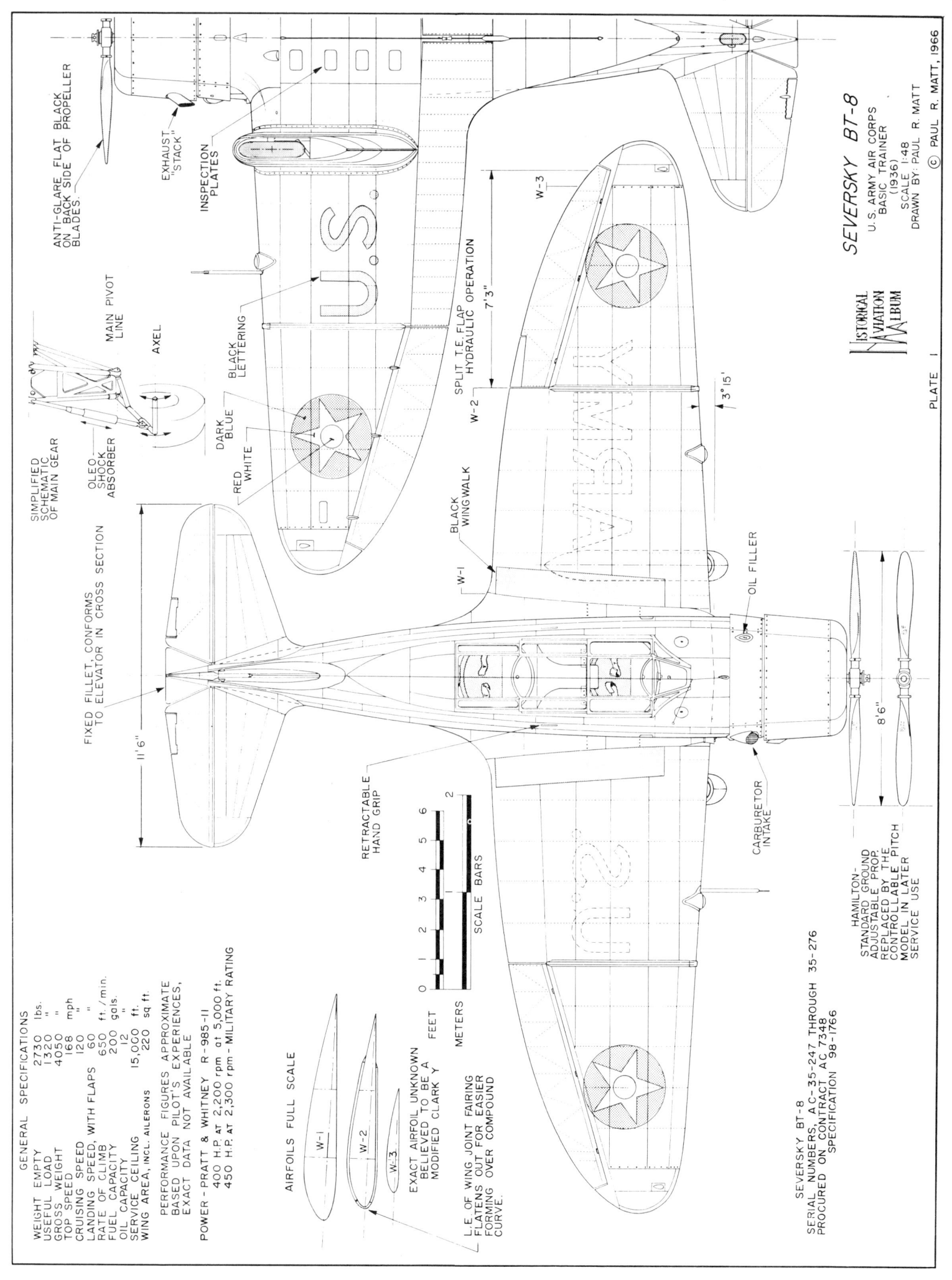

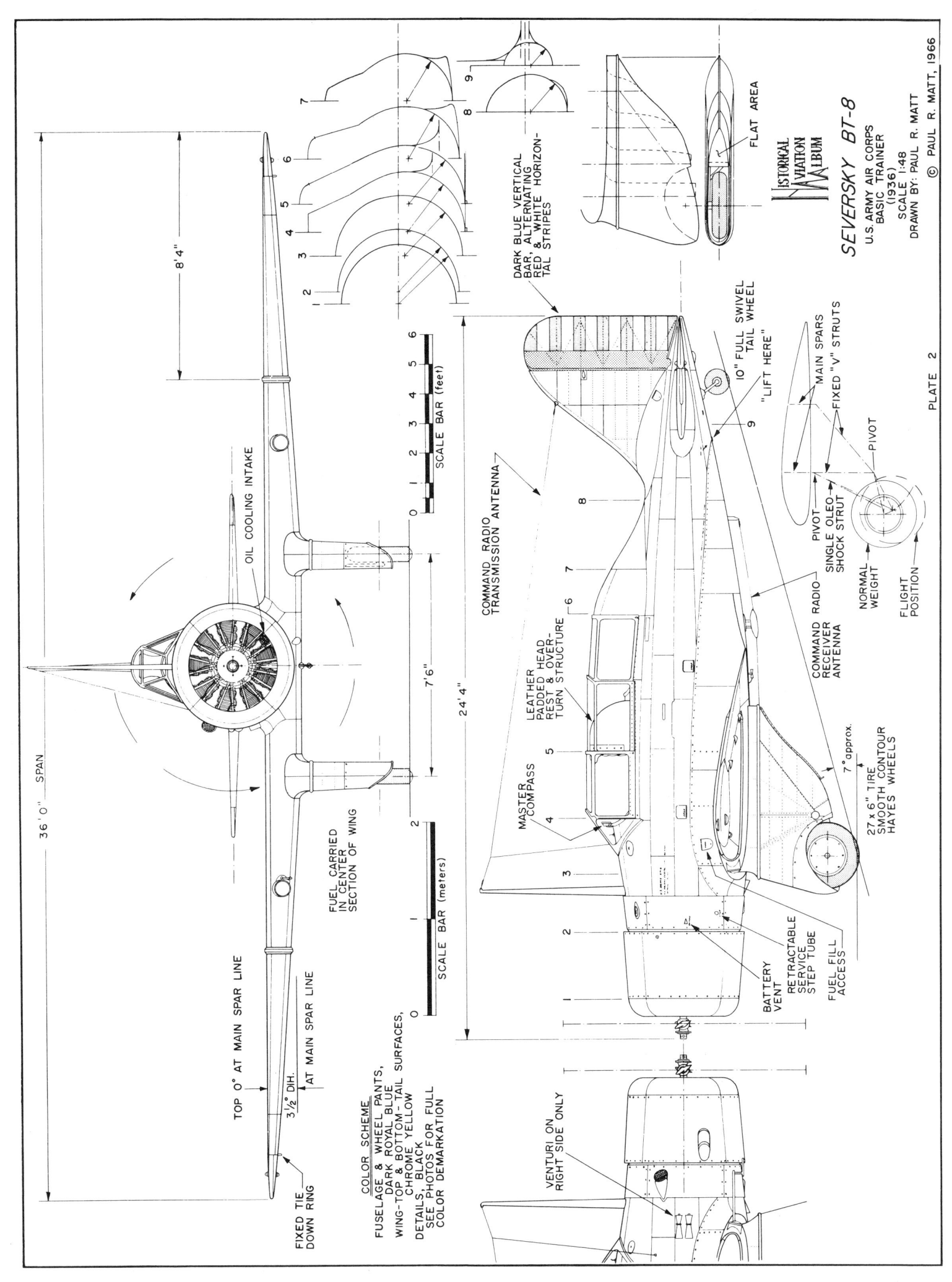

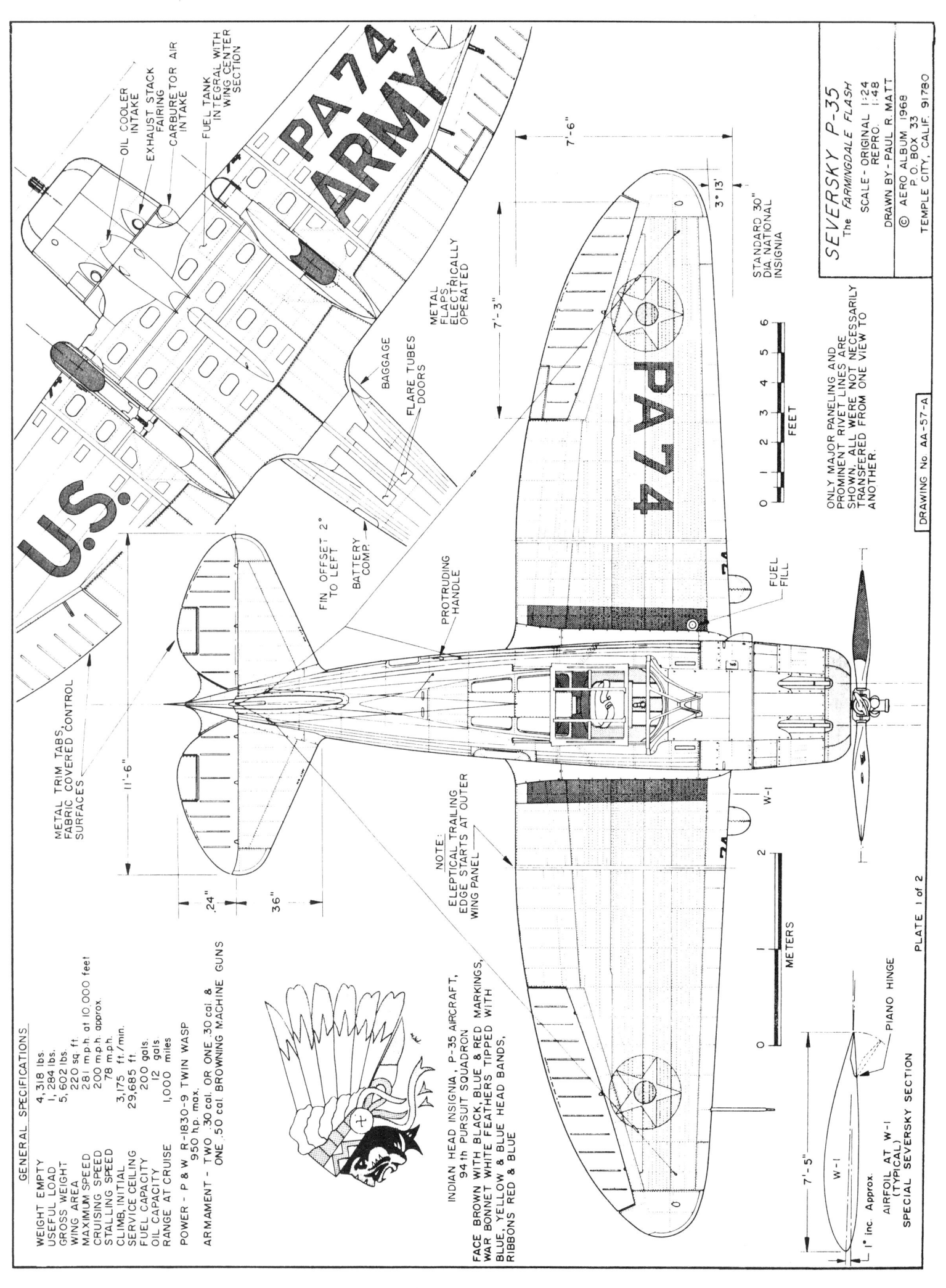

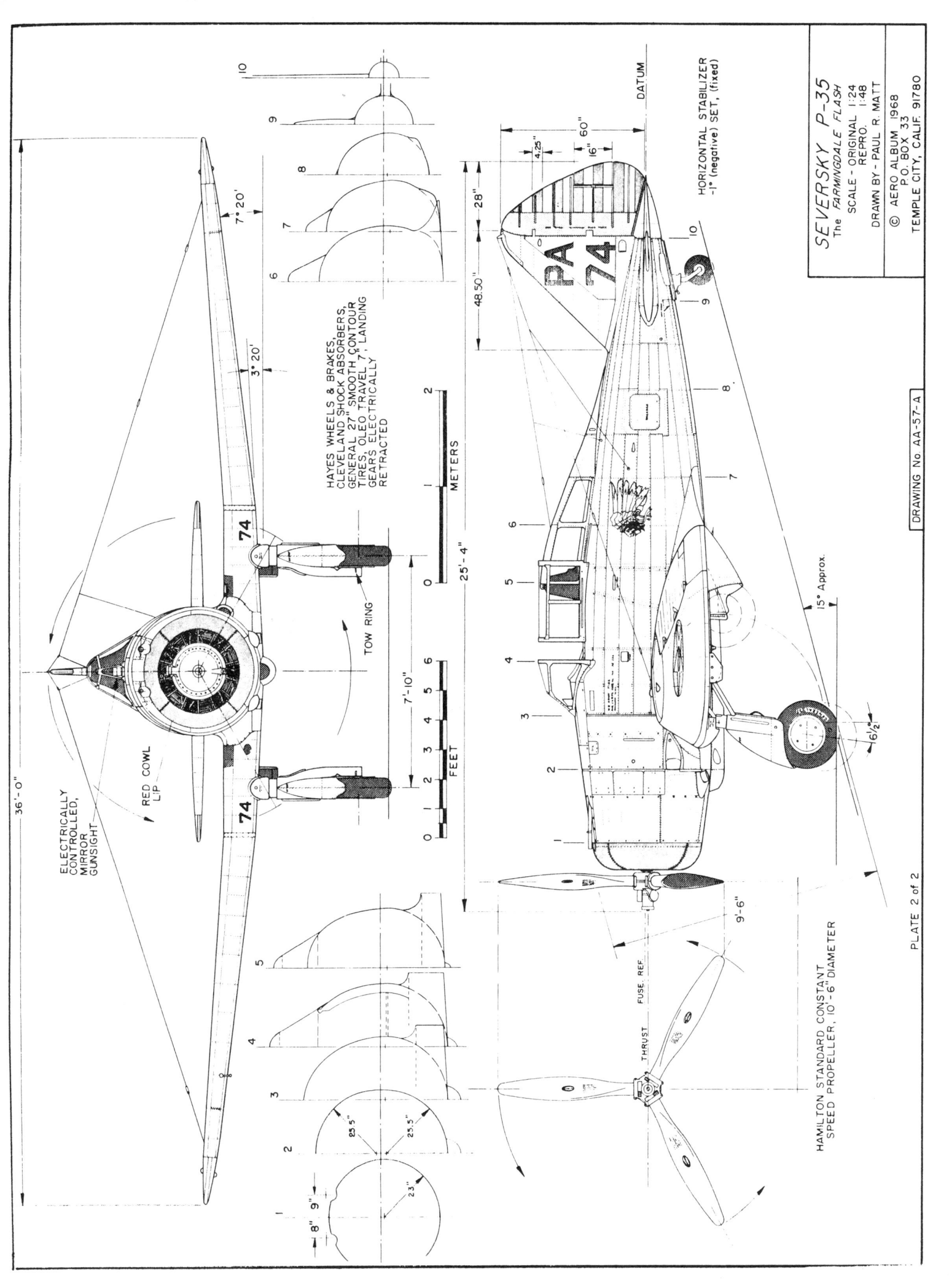

Sikorsky S-39B

Sikorsky S-39B instrument panel.

Sikorsky S-39B interior.

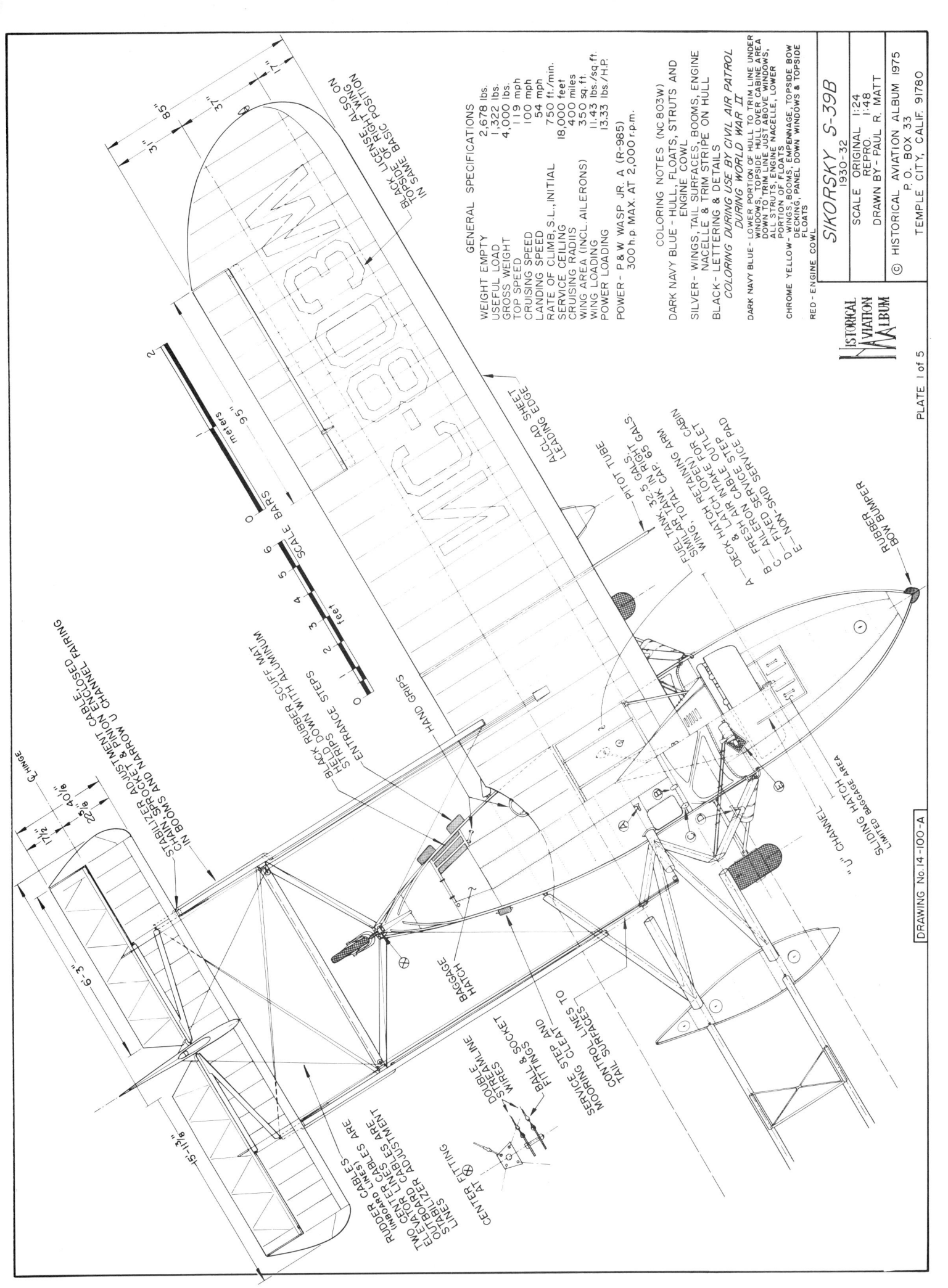

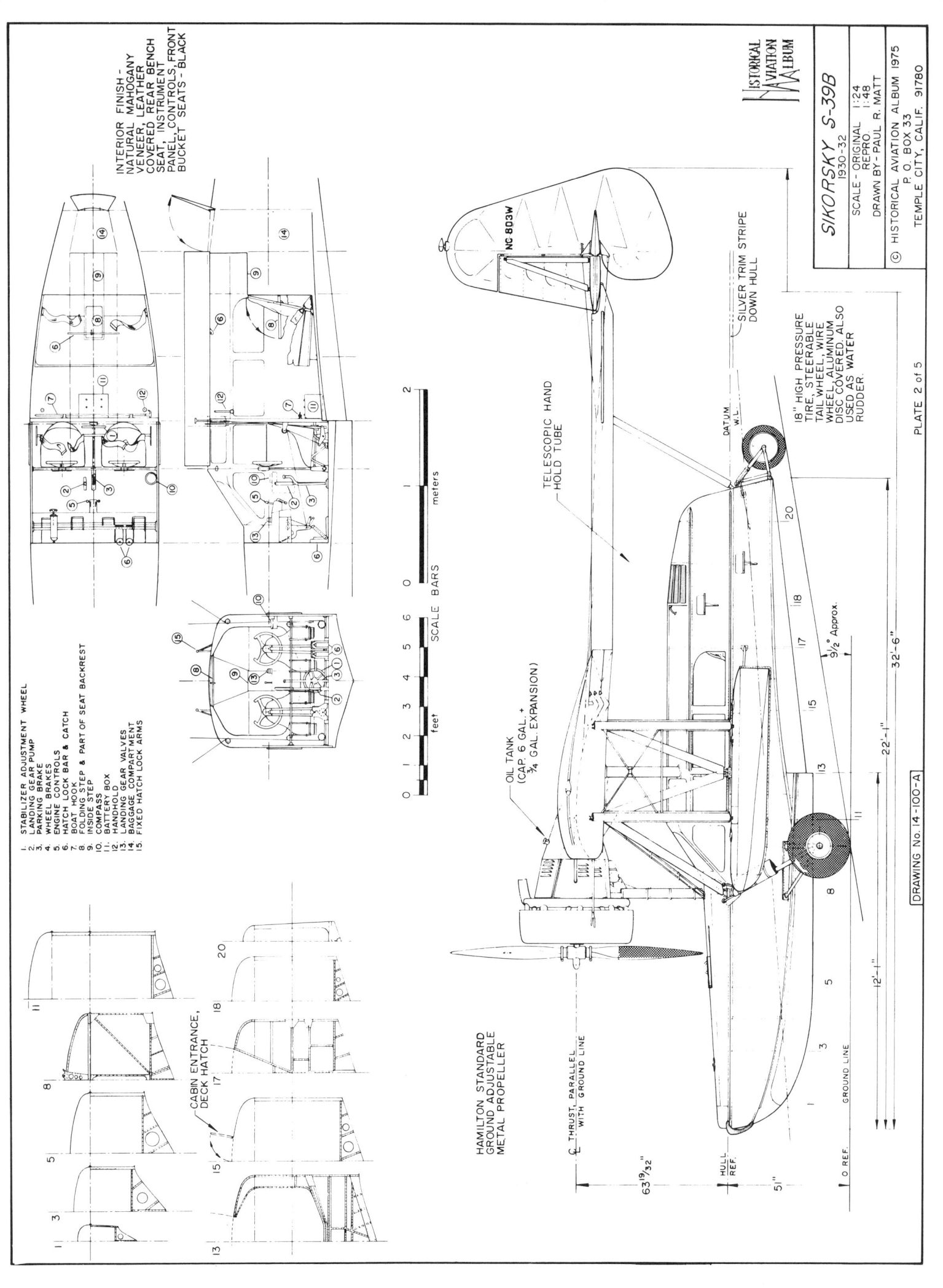

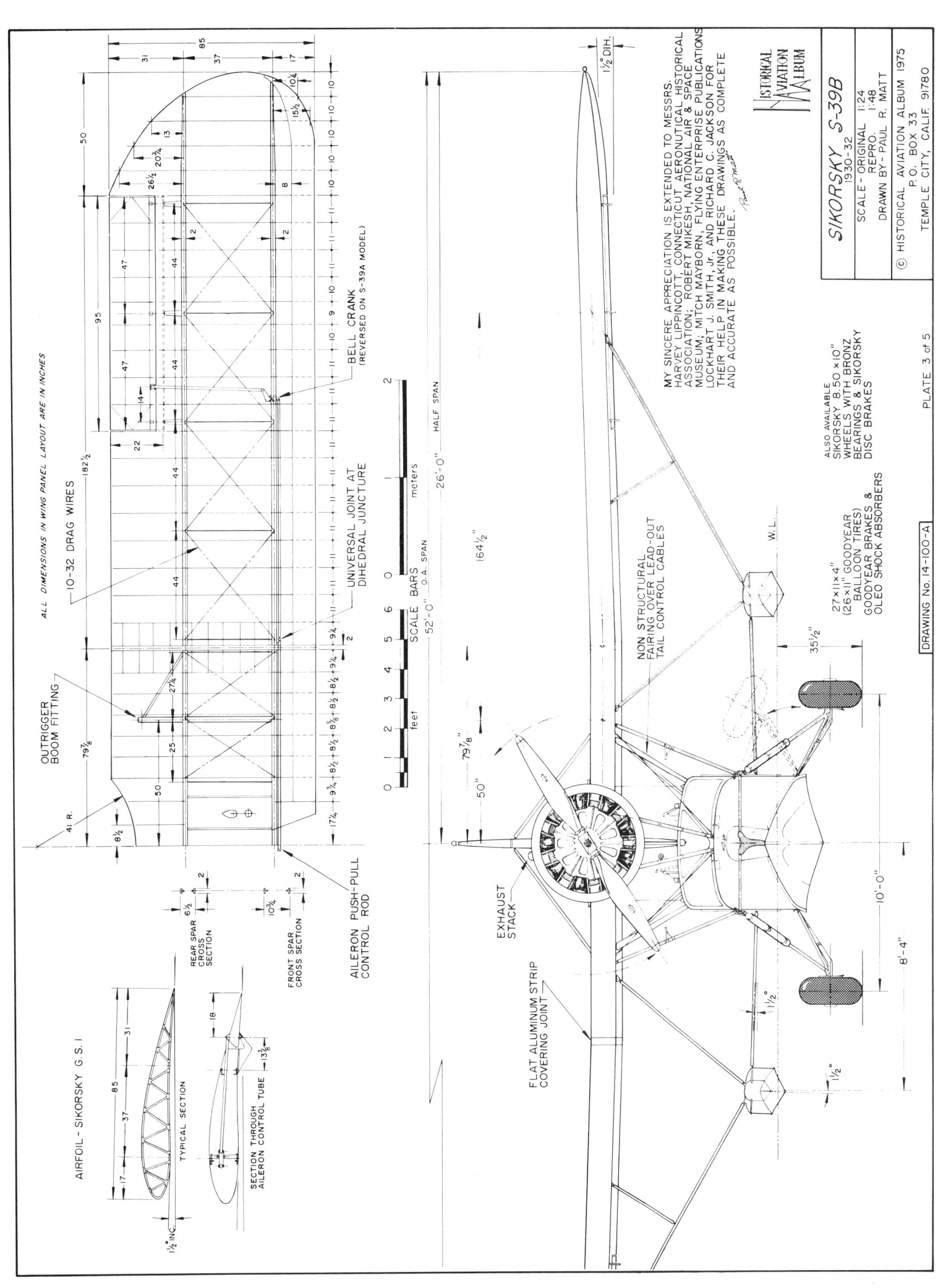

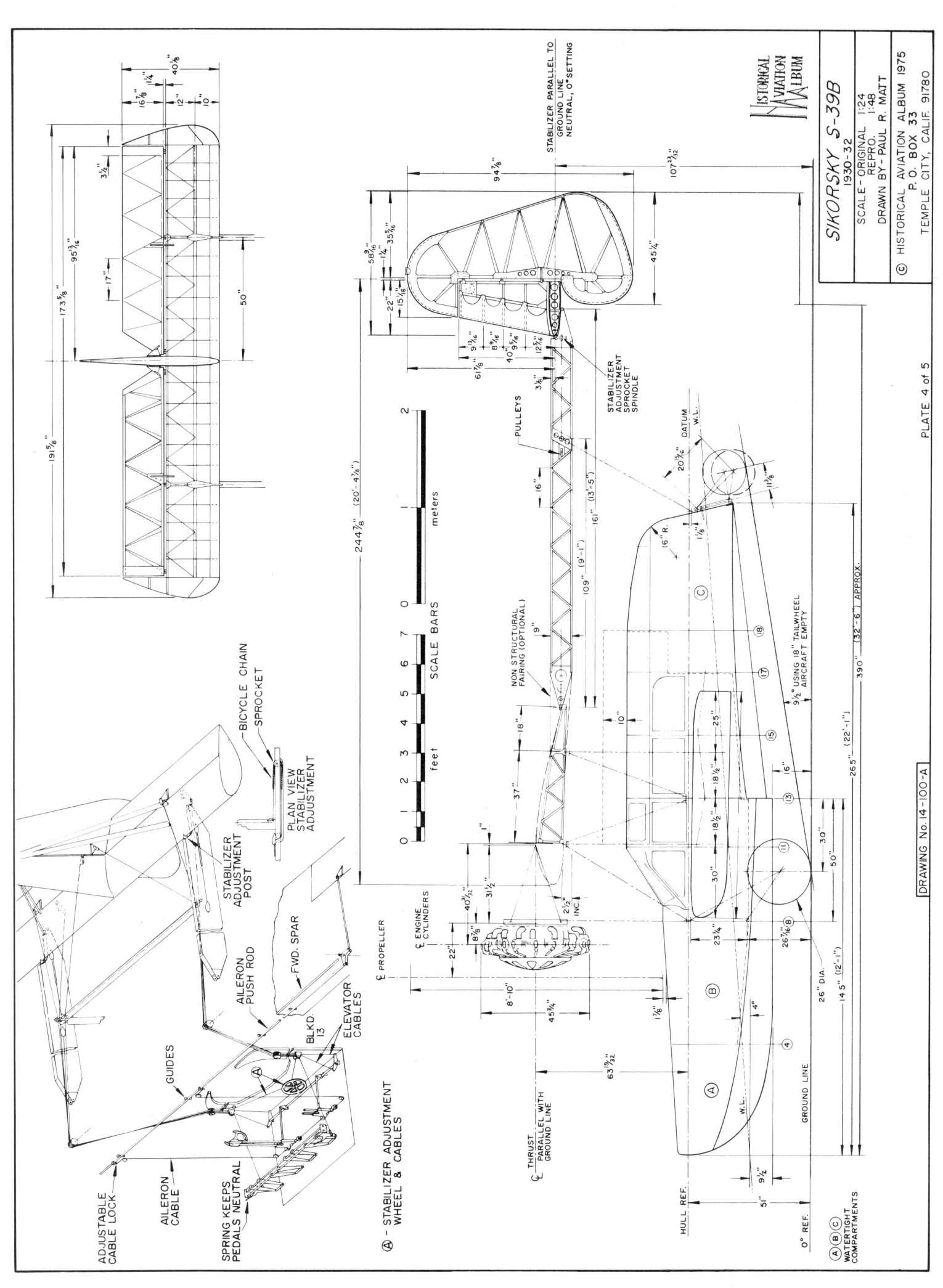

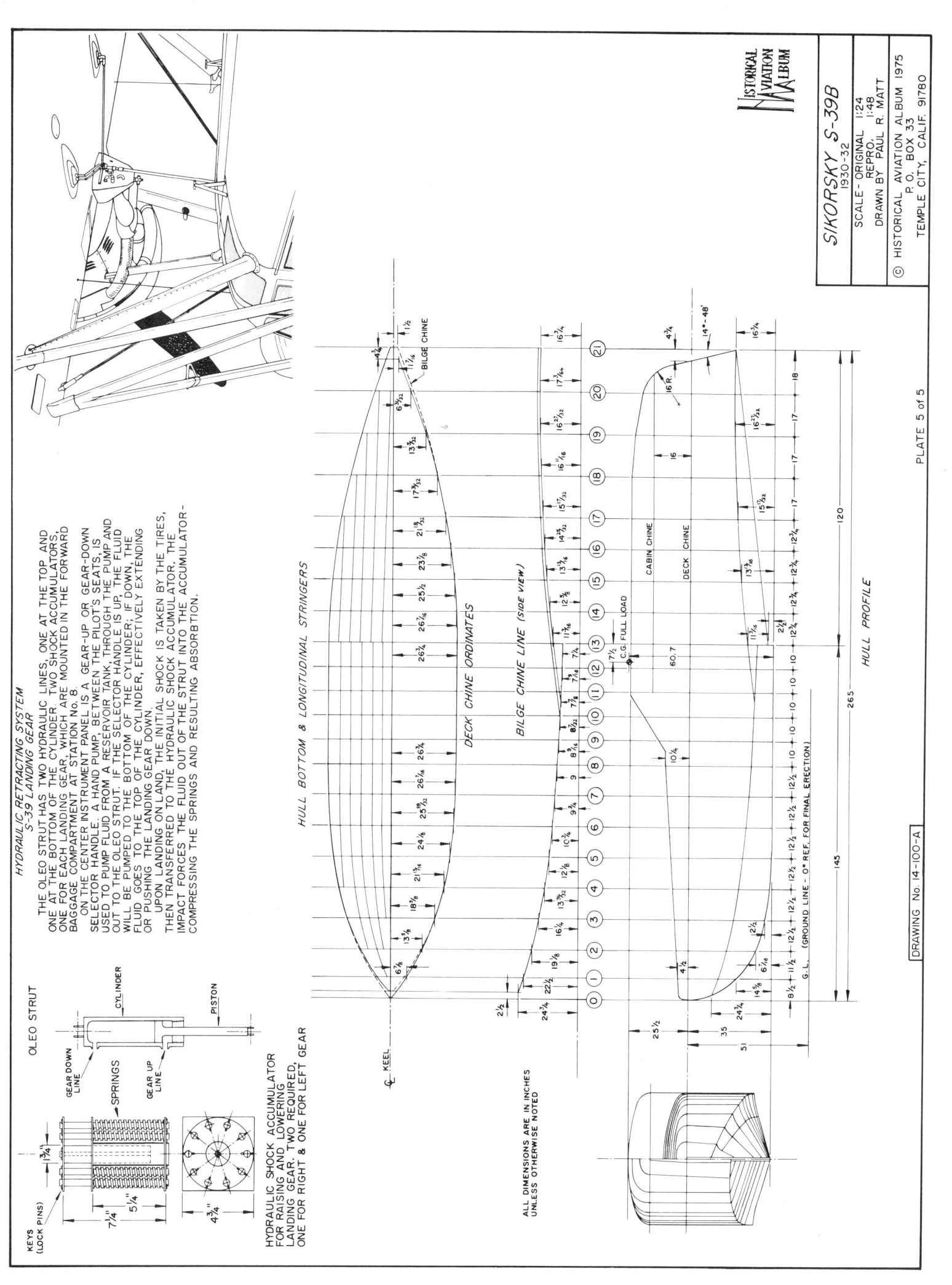

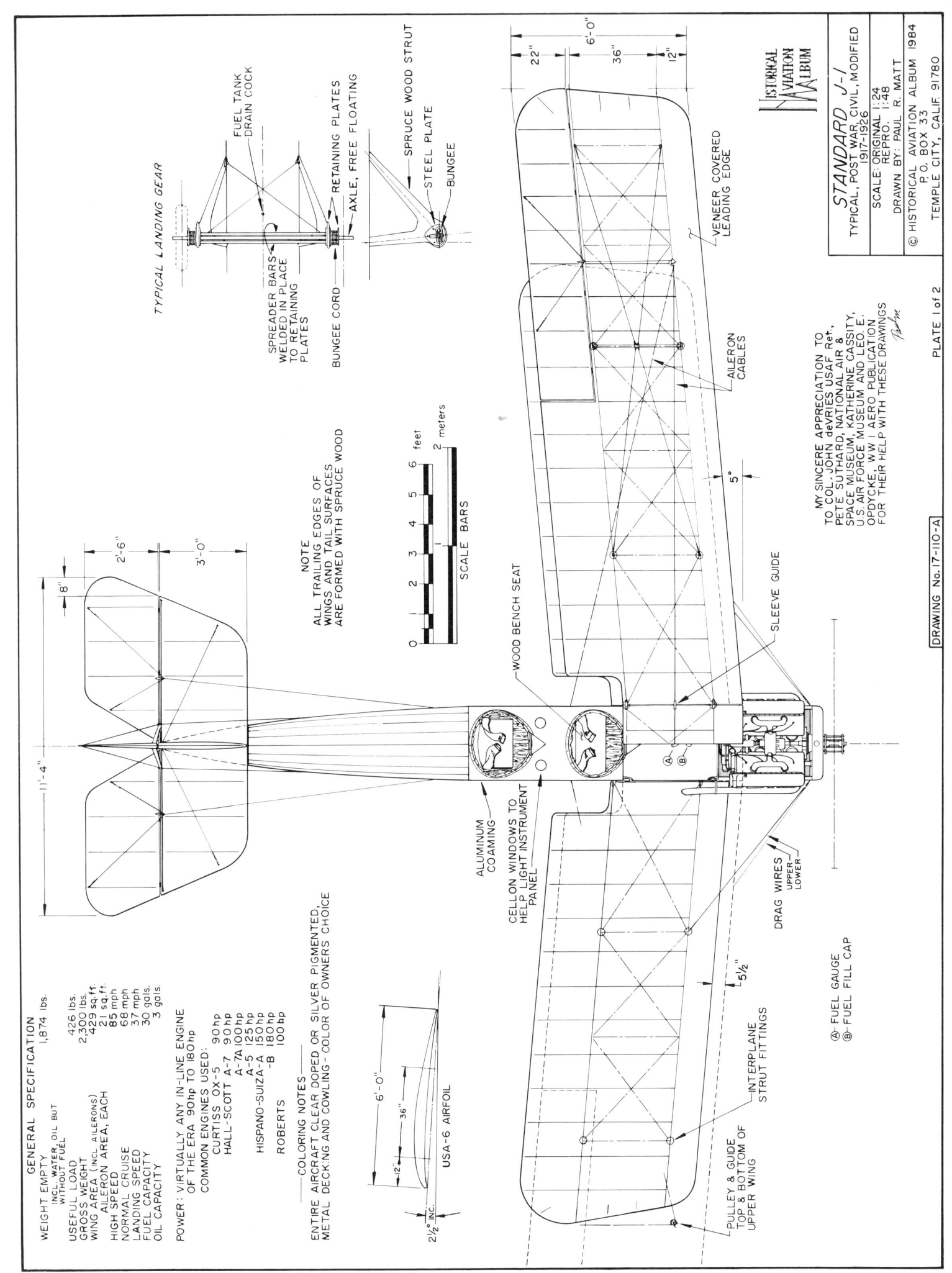

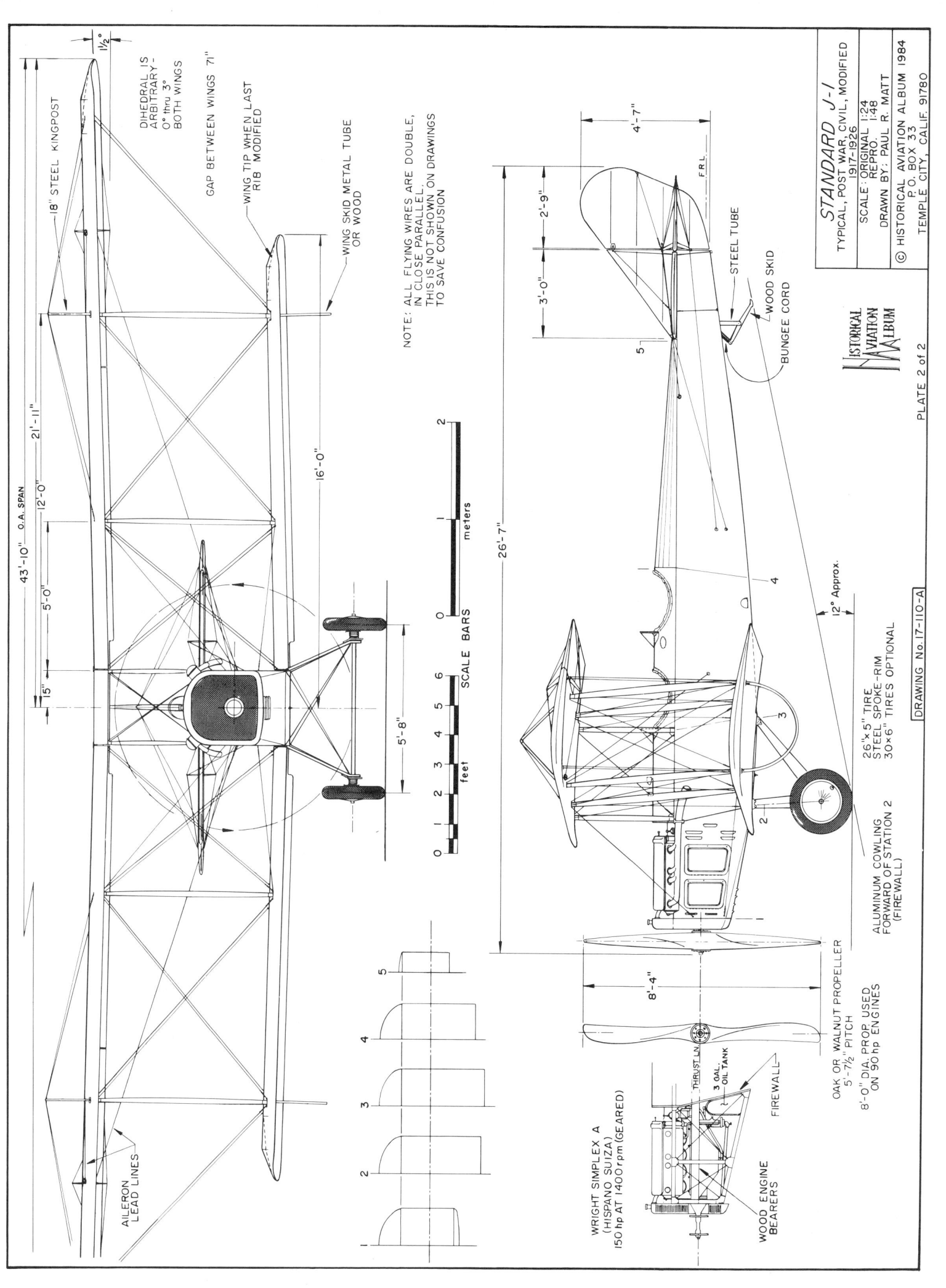

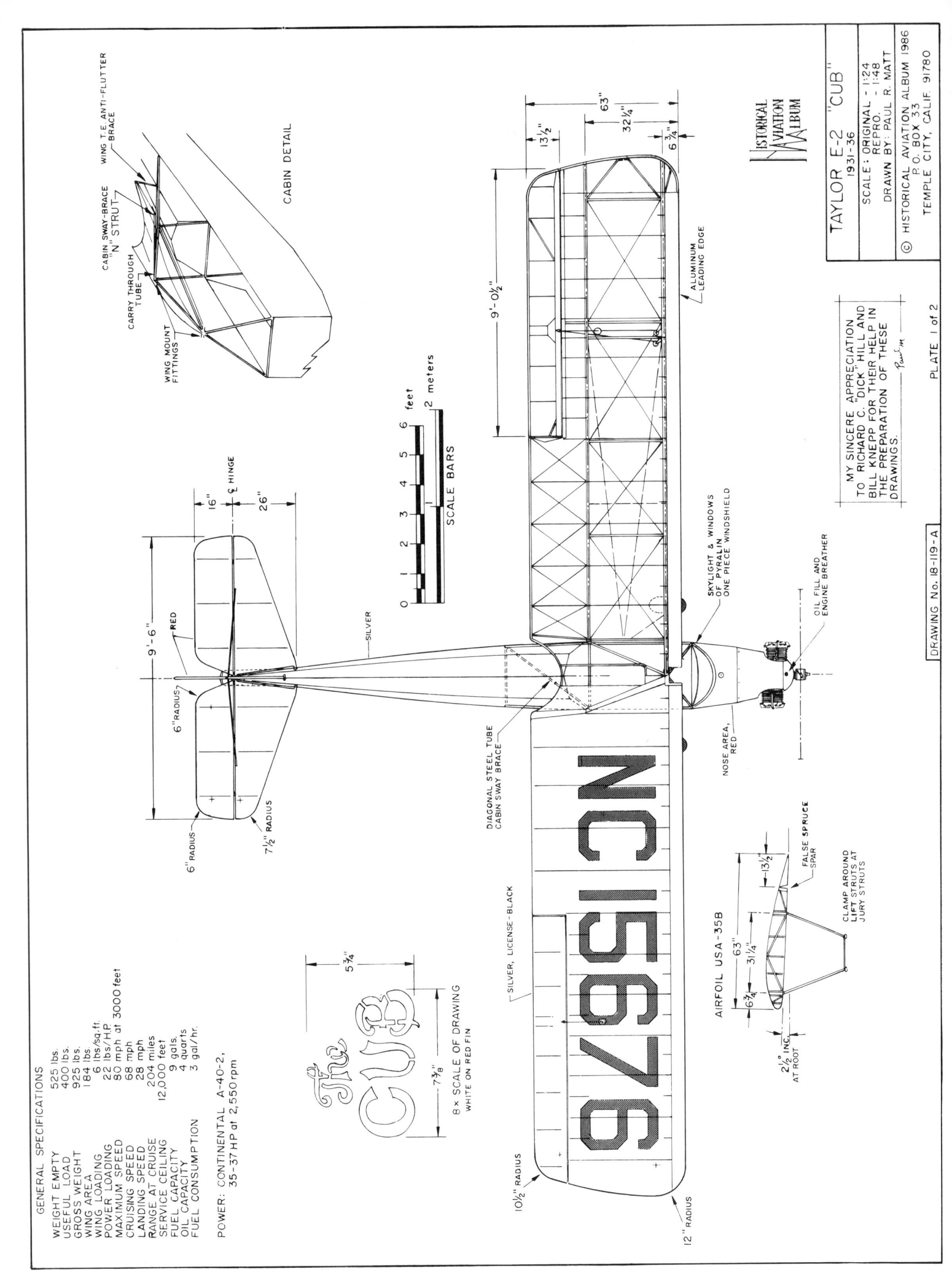

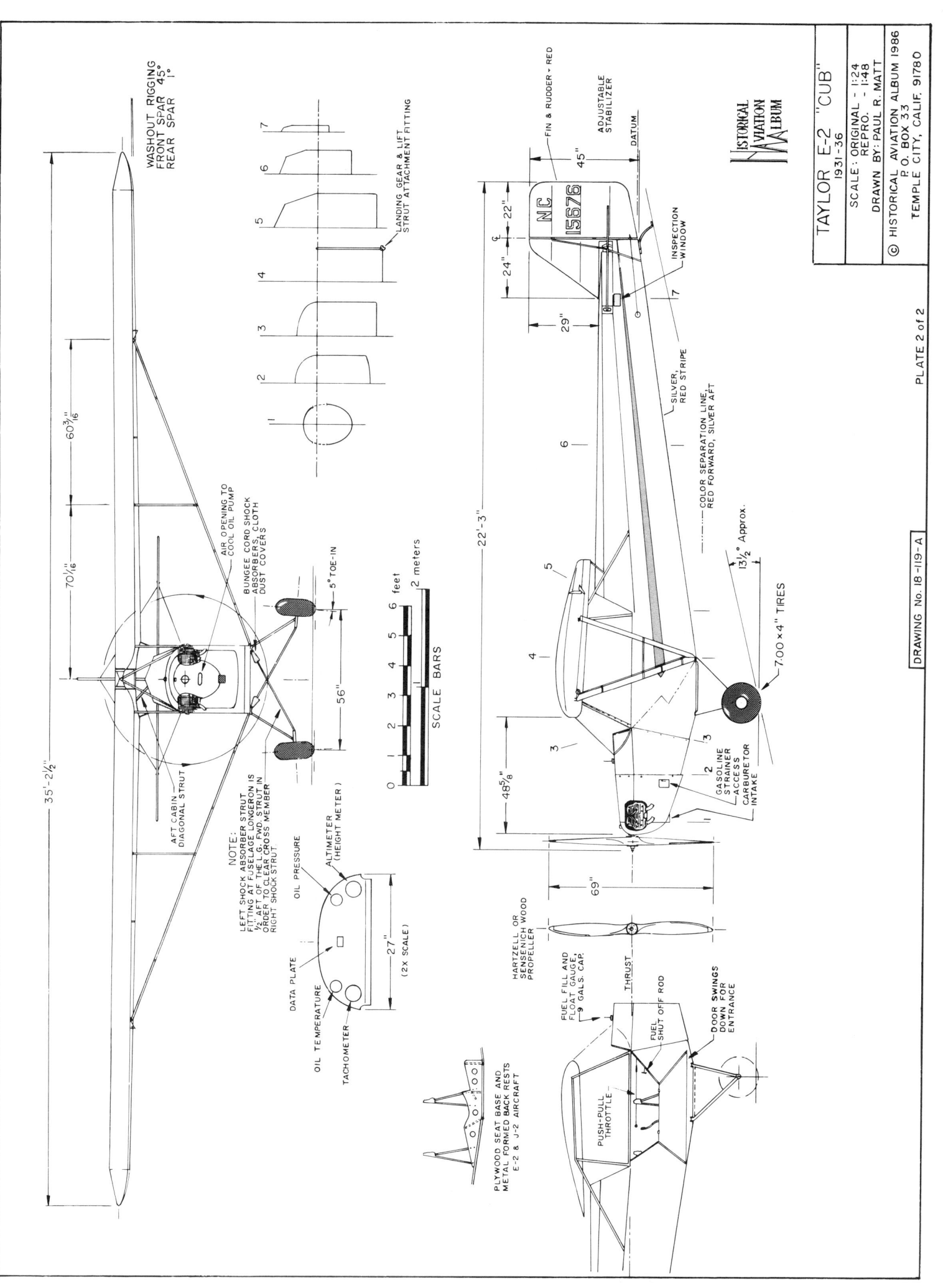

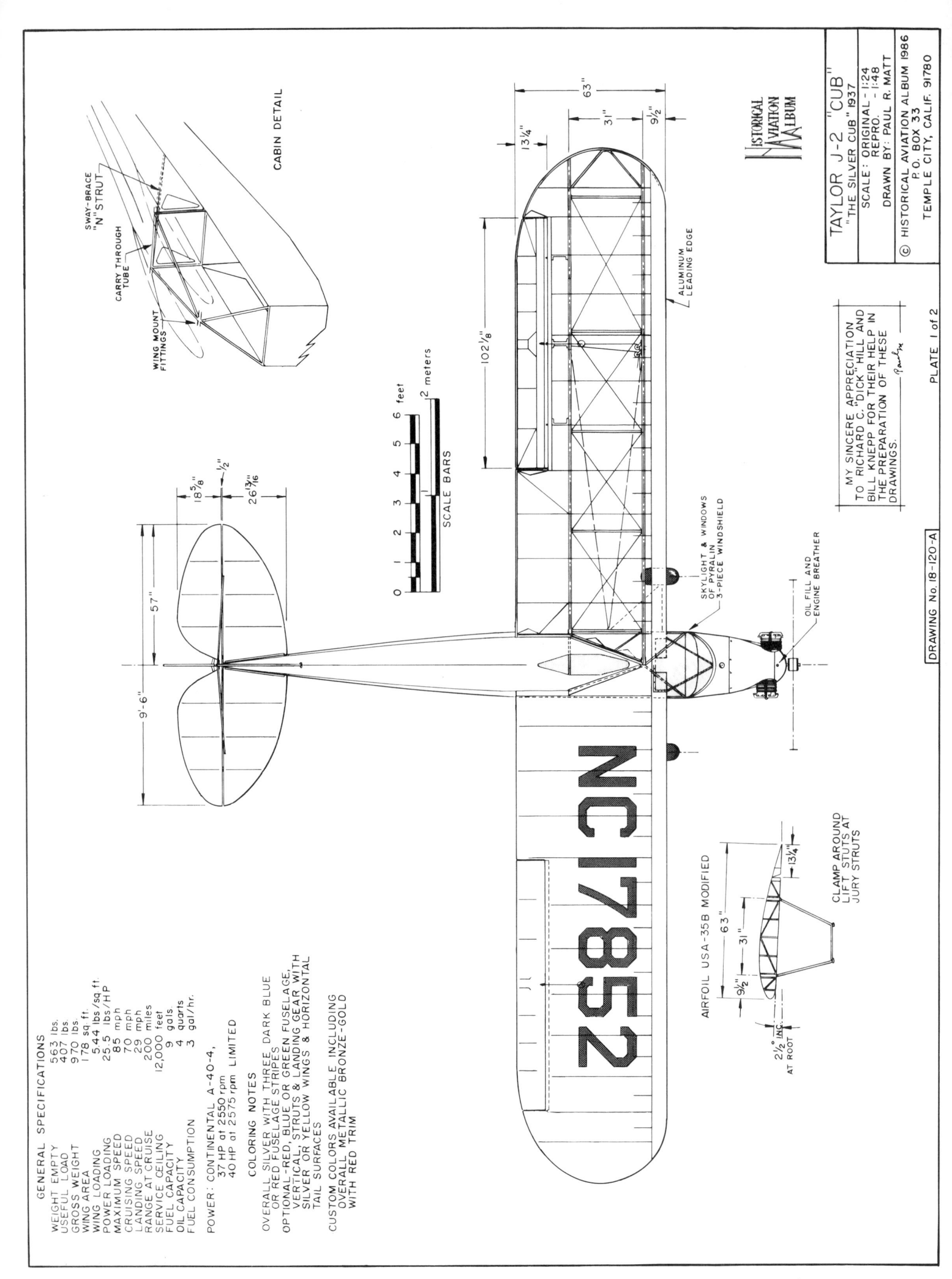

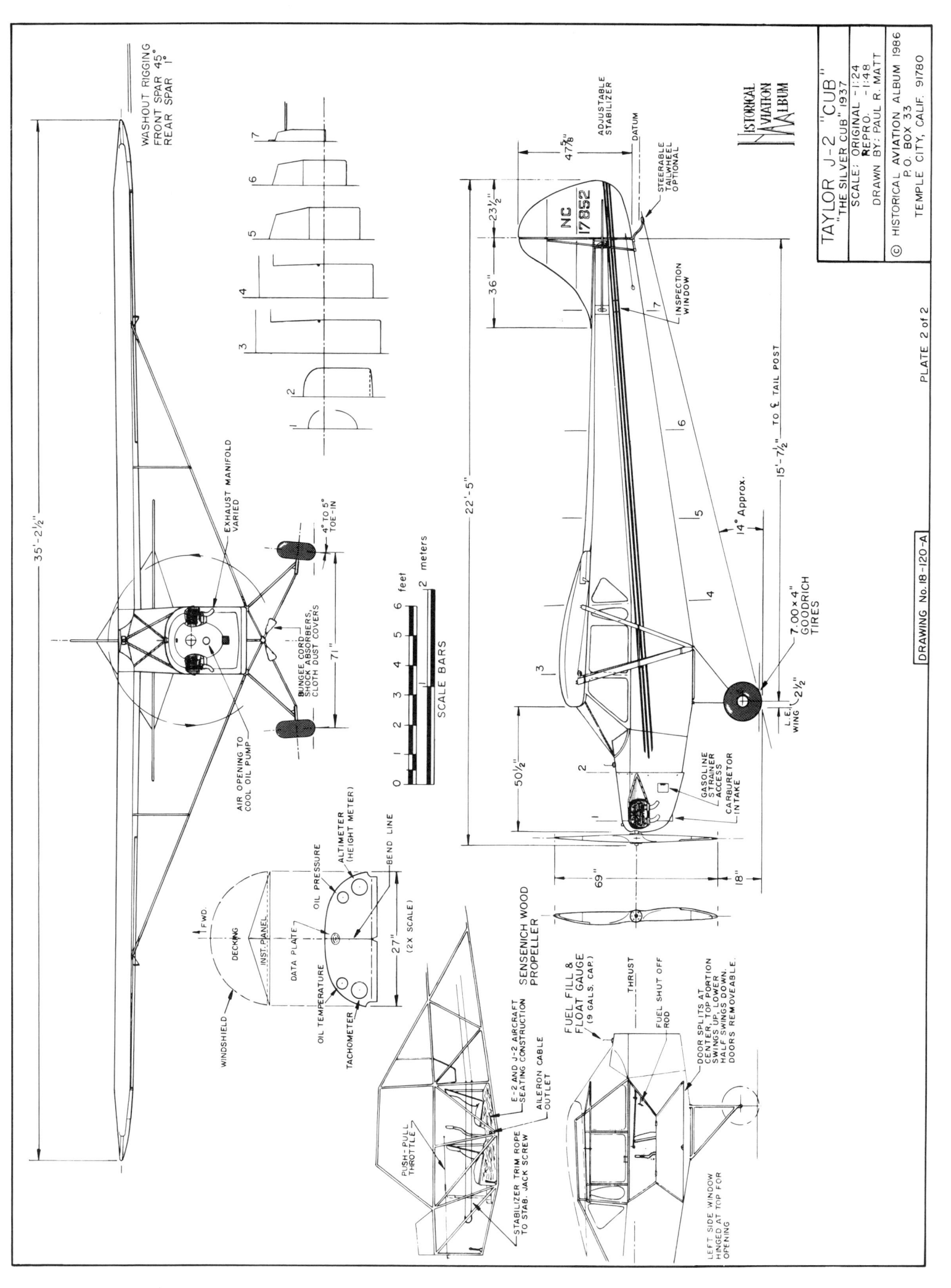

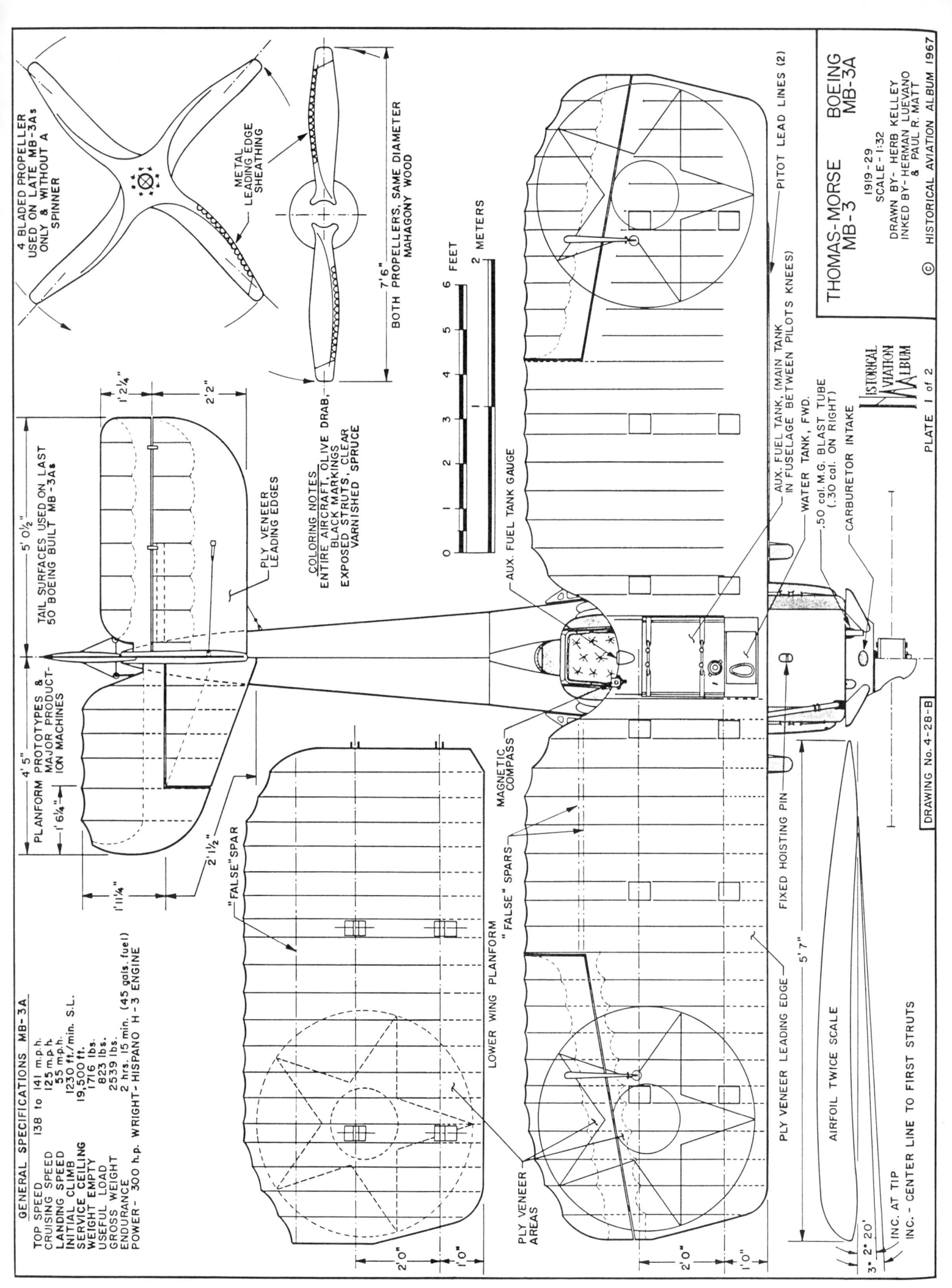

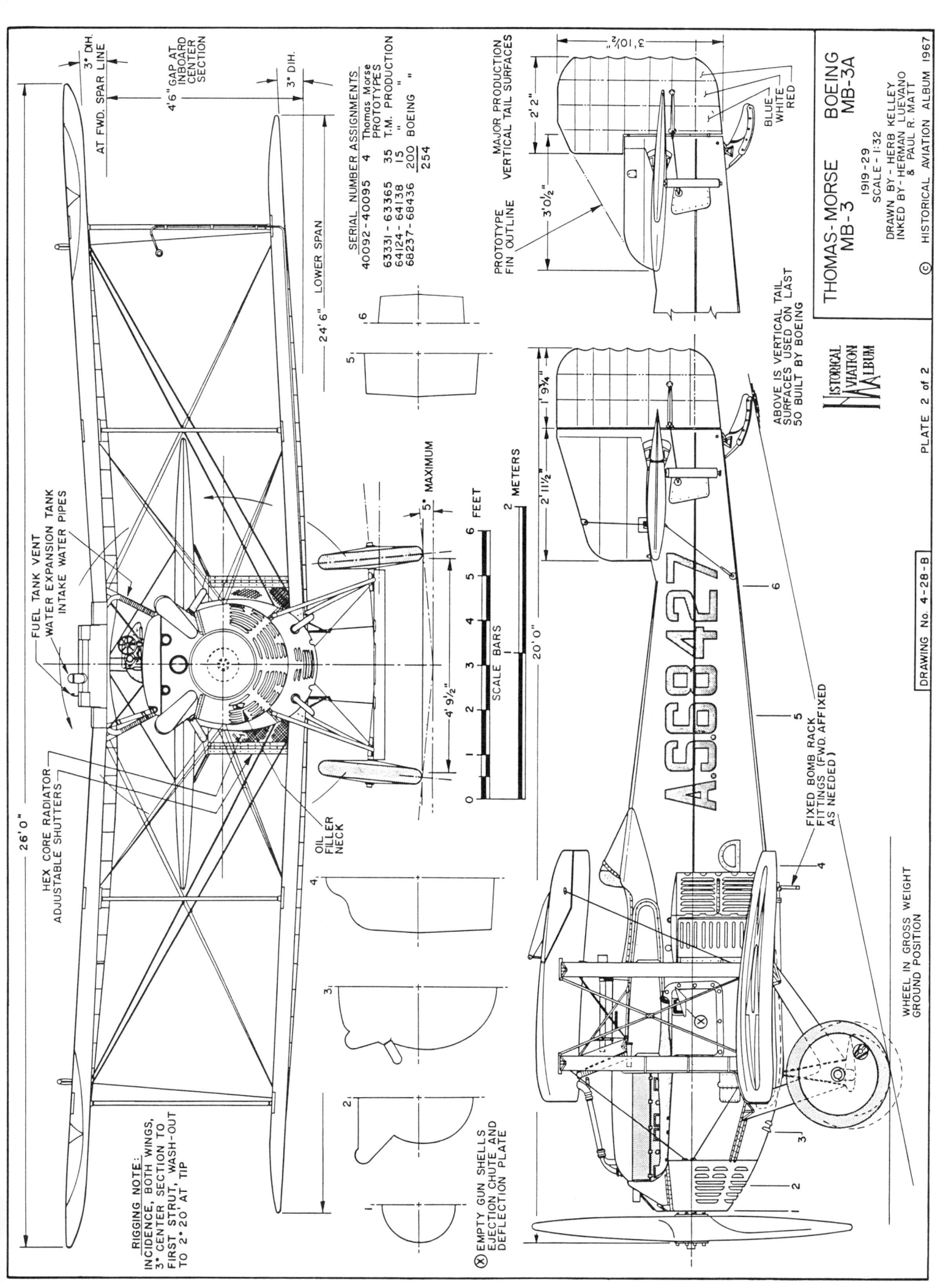

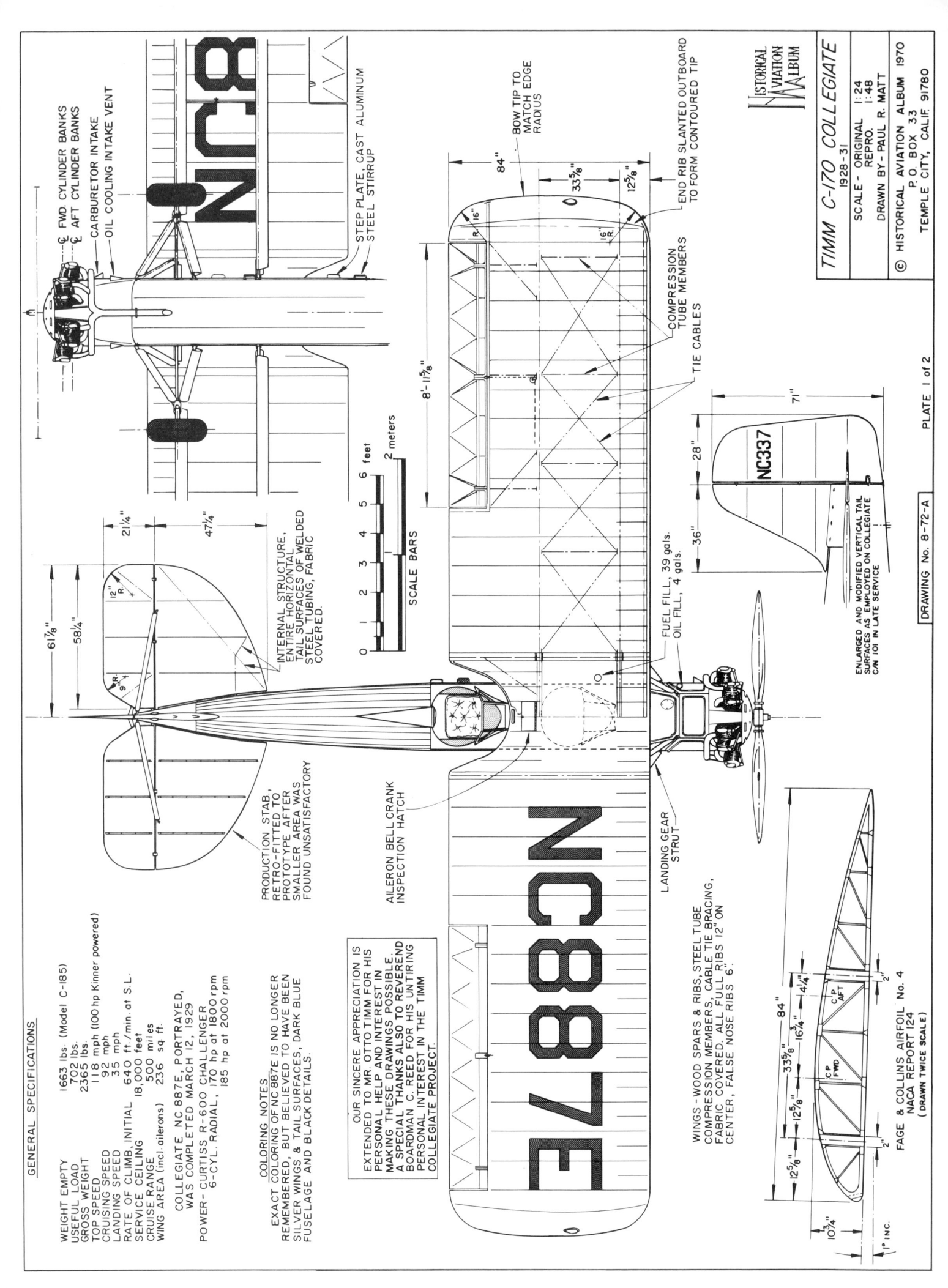

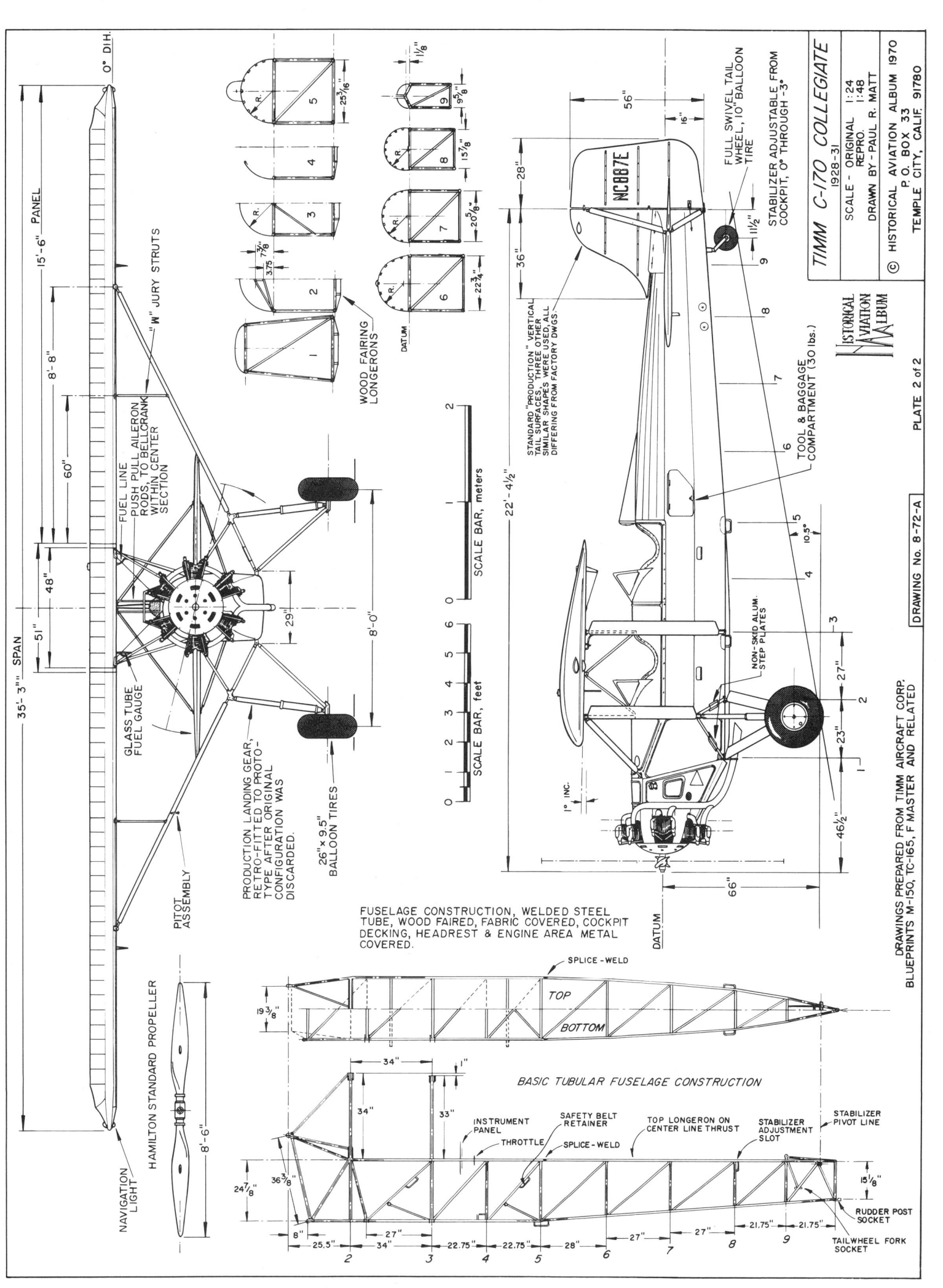

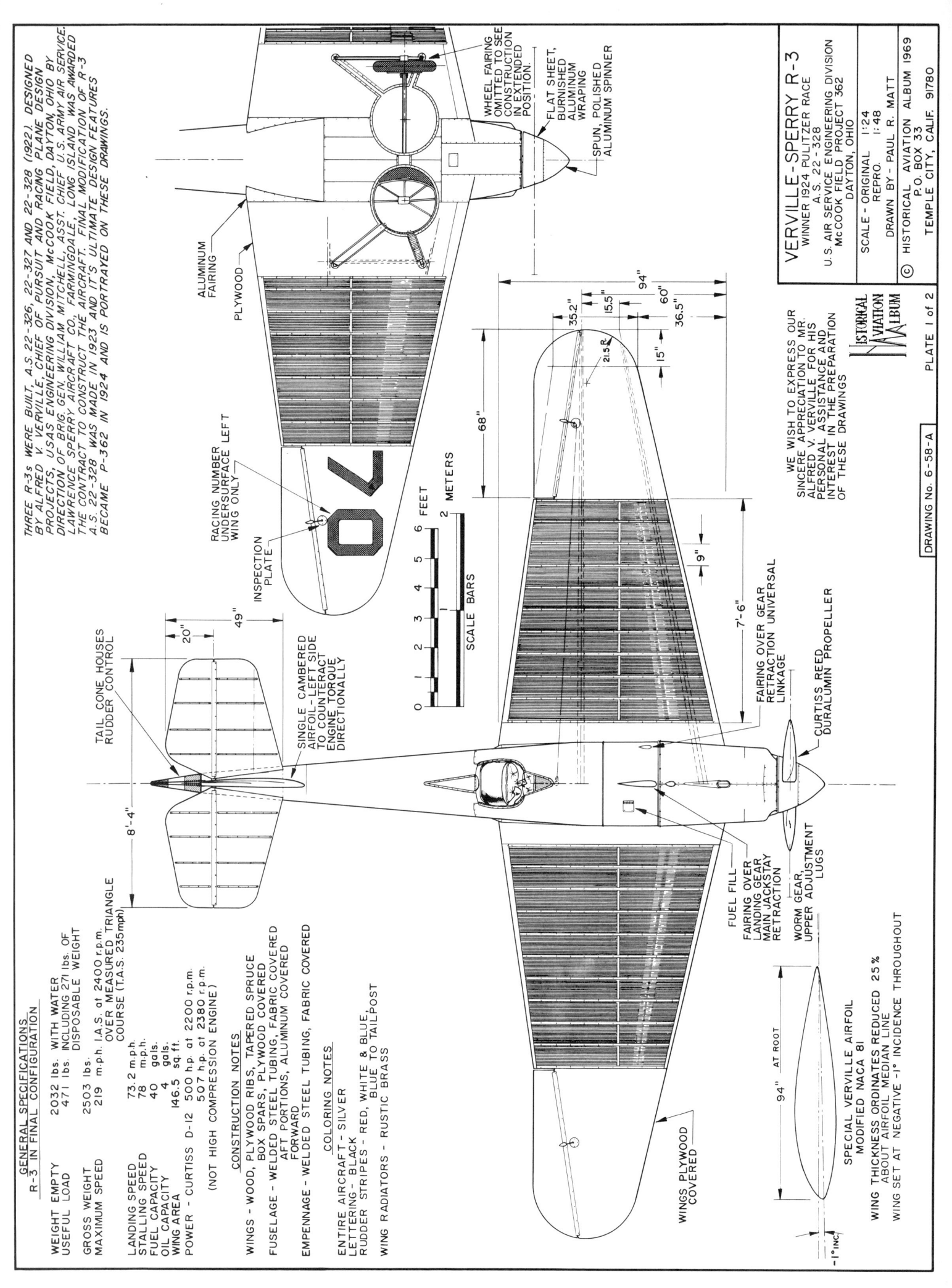

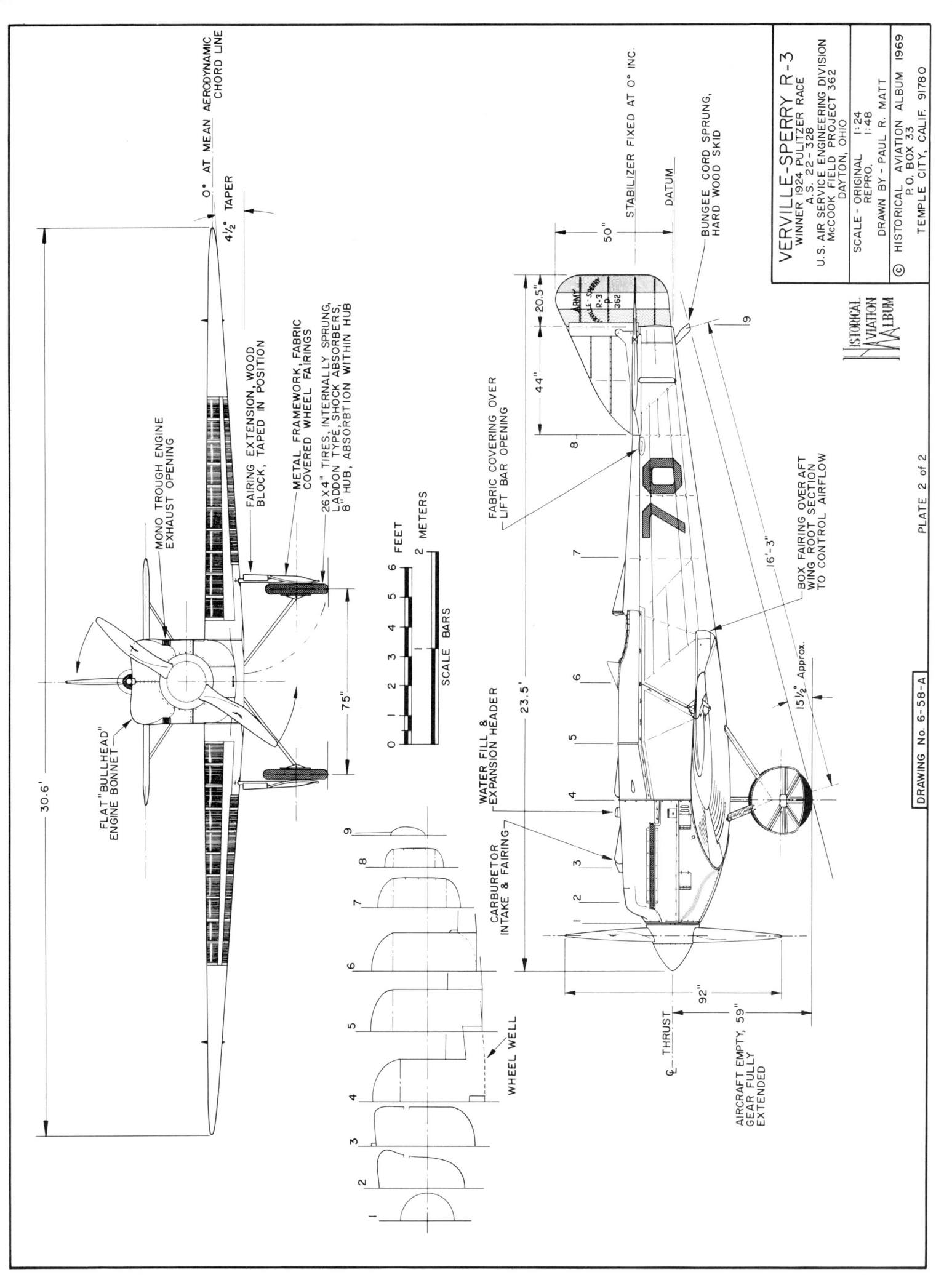

Vought F4U-1 "Corsair"

F4U-1D "Corsair", June 6, 1944.

Goodyear FG-1D "Corsair".

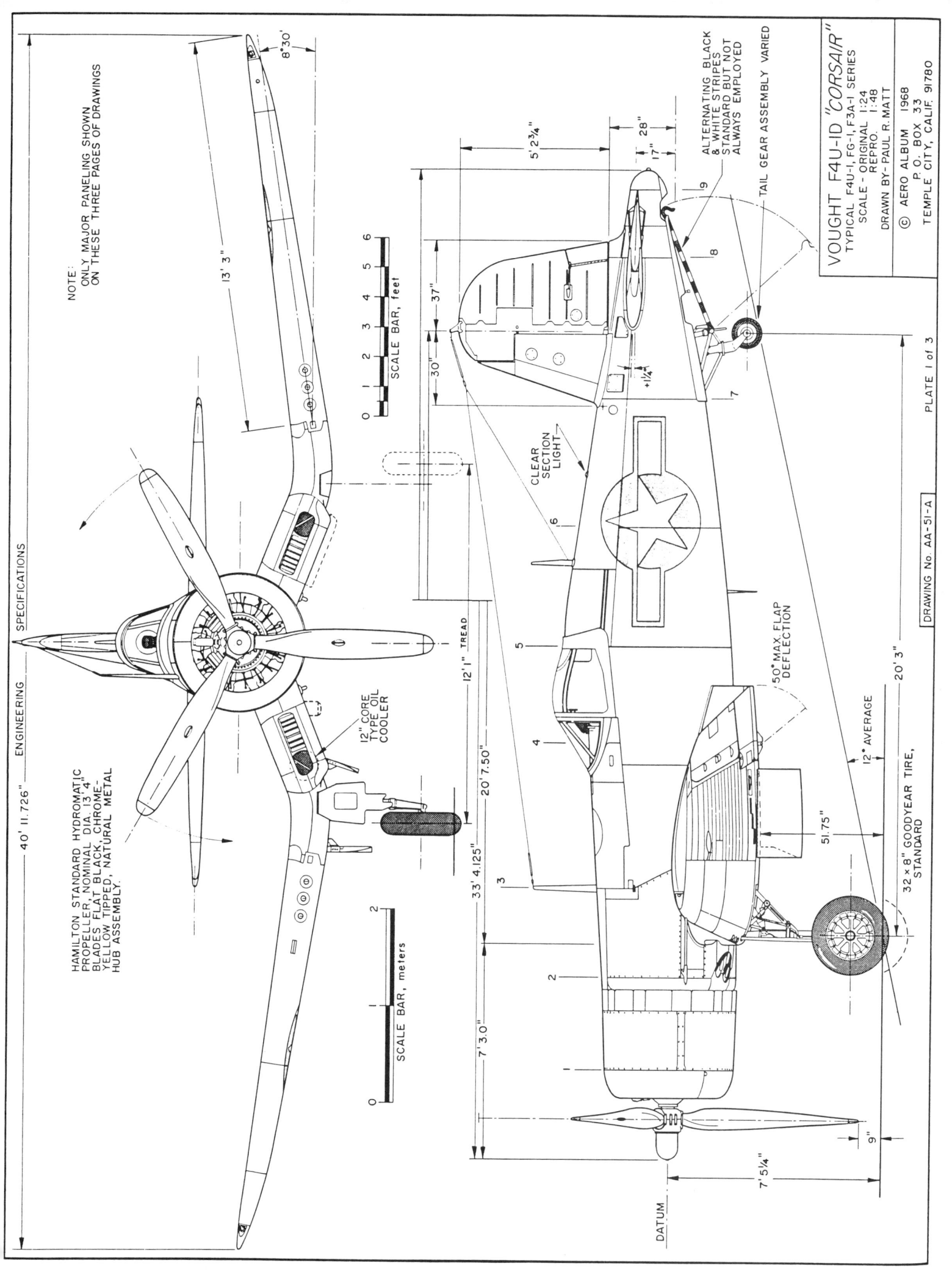

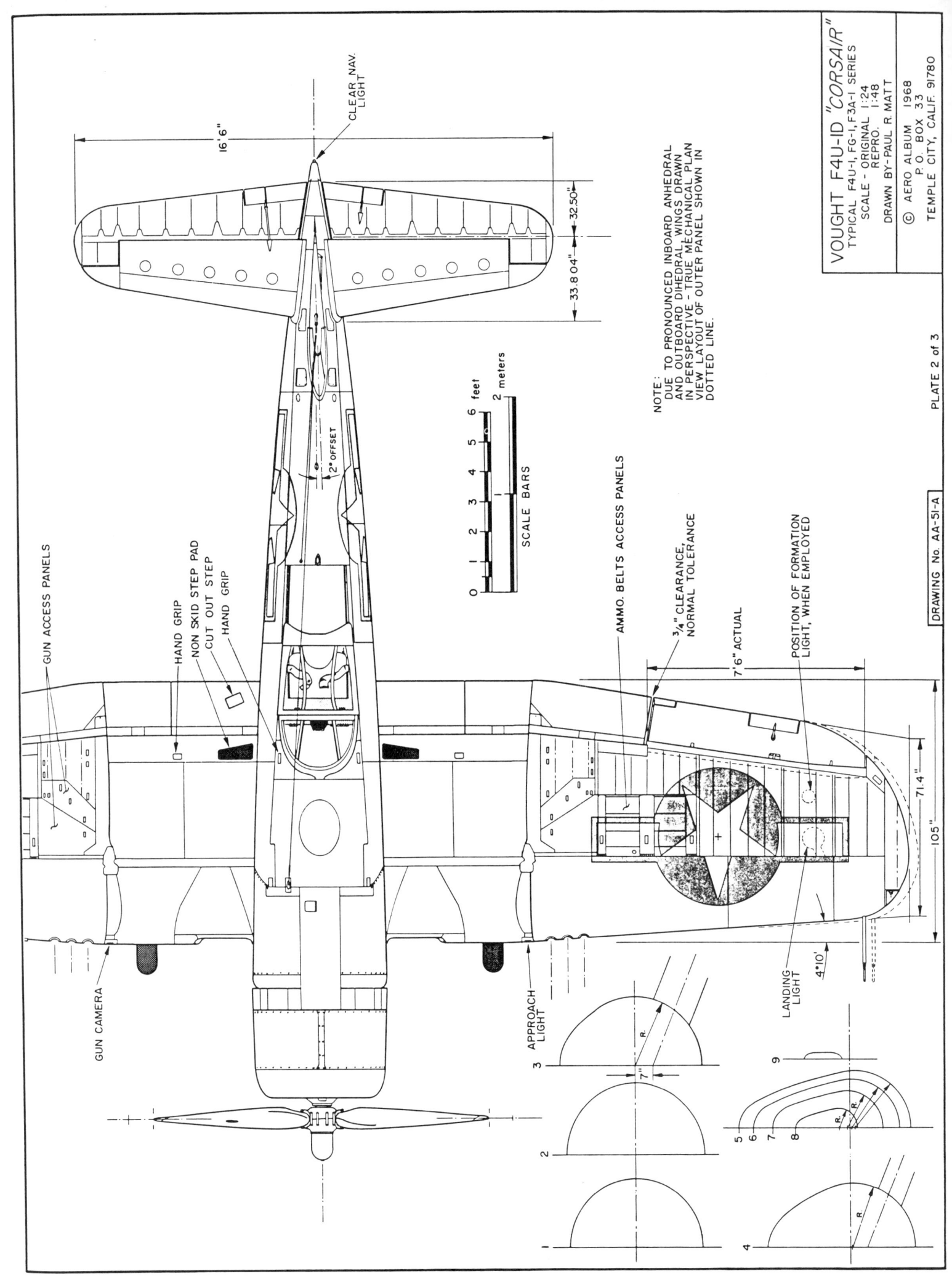

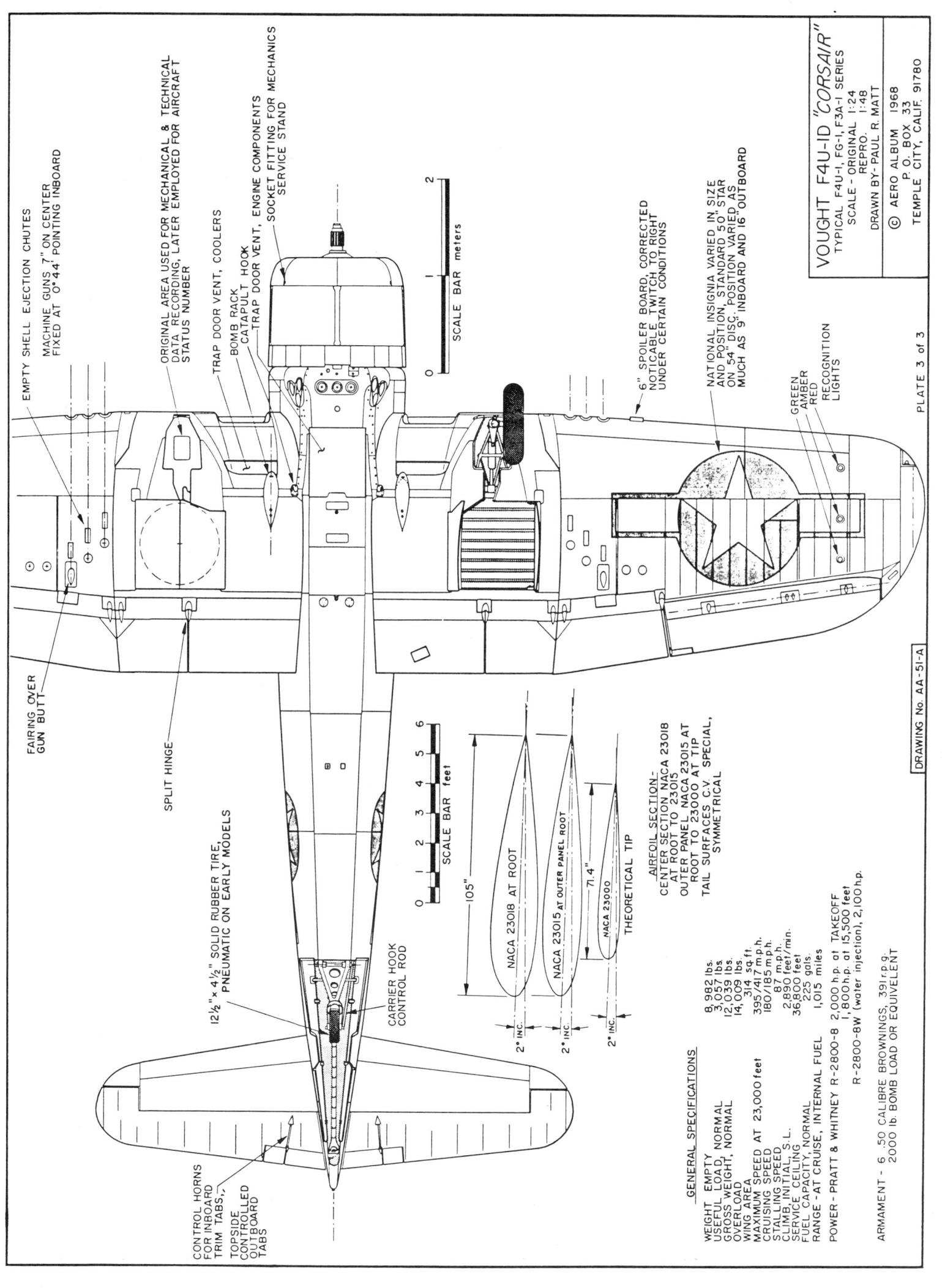

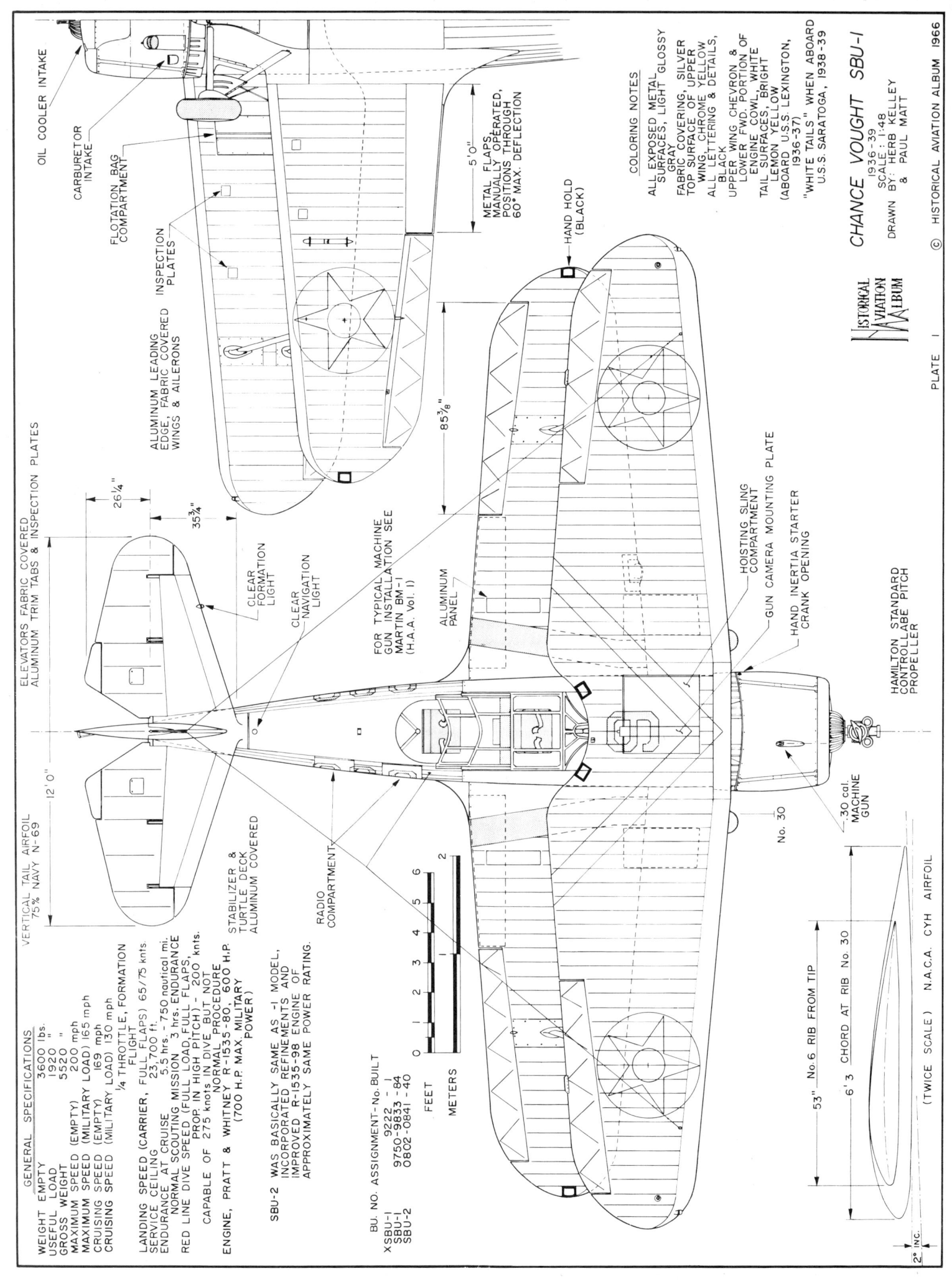

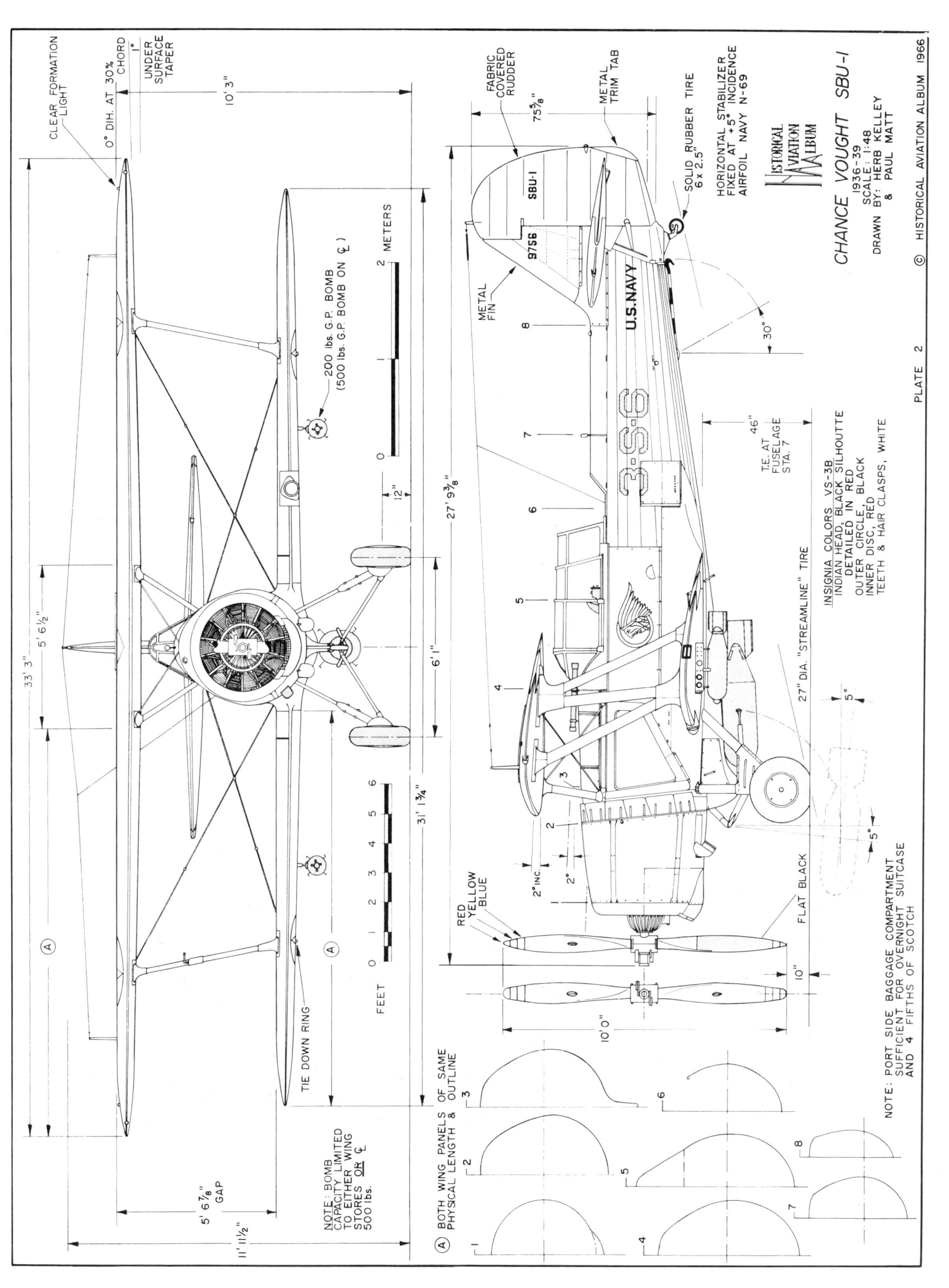

Vought XF5U-1

Vought XF5U-1, mock up completed.

Vought XF5U-1 airplane under ground run conditions, August 21, 1947.

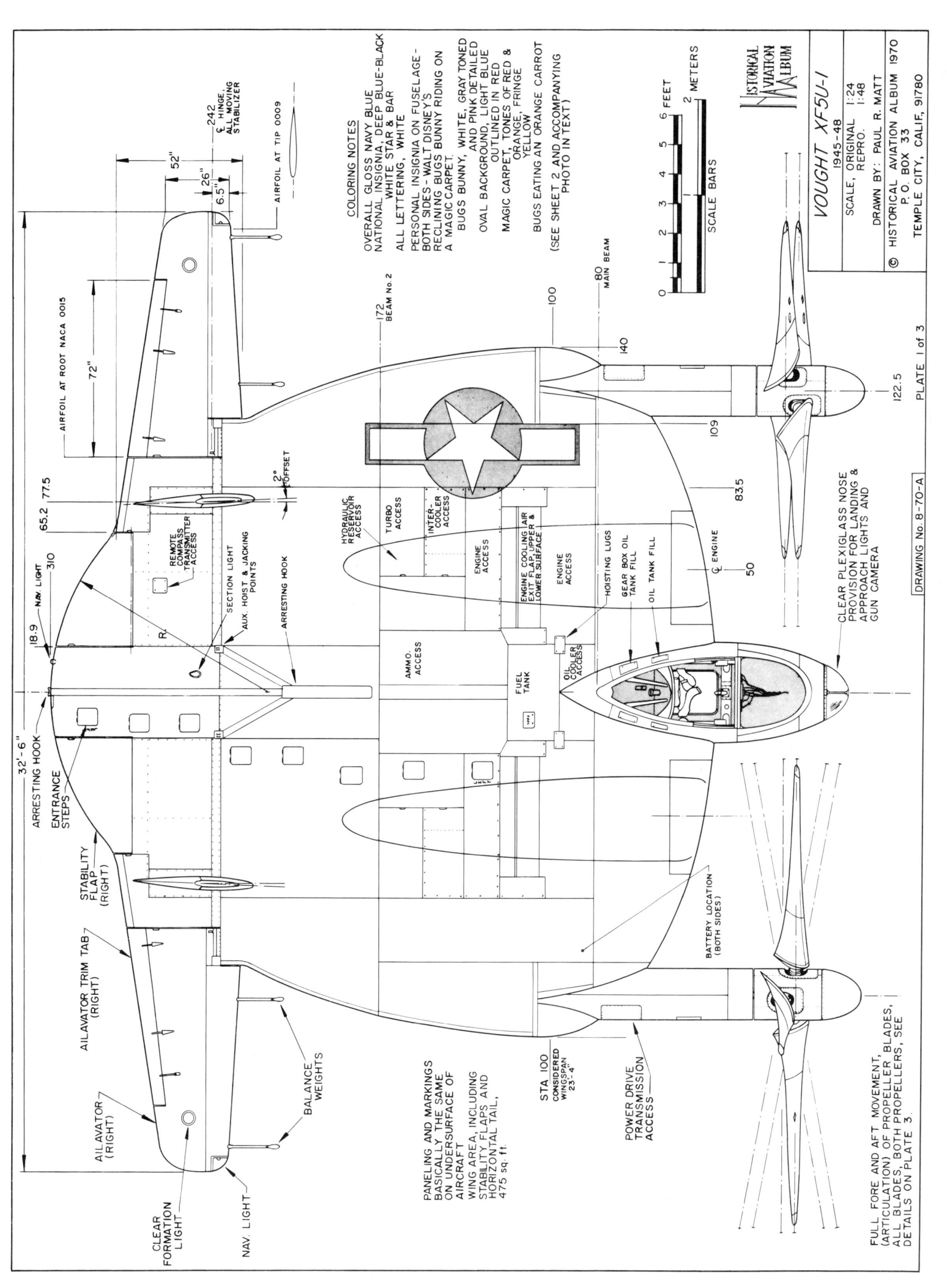

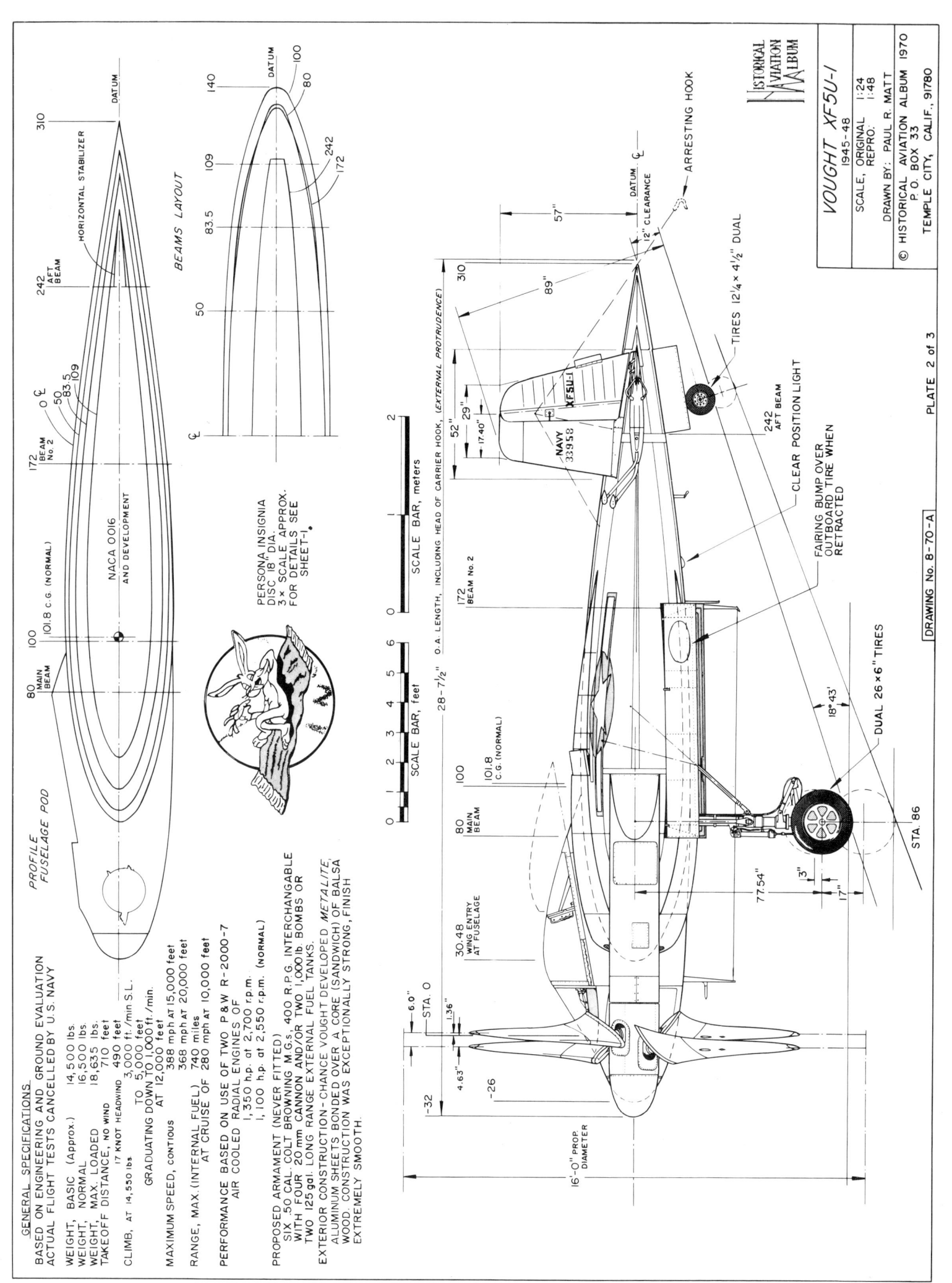

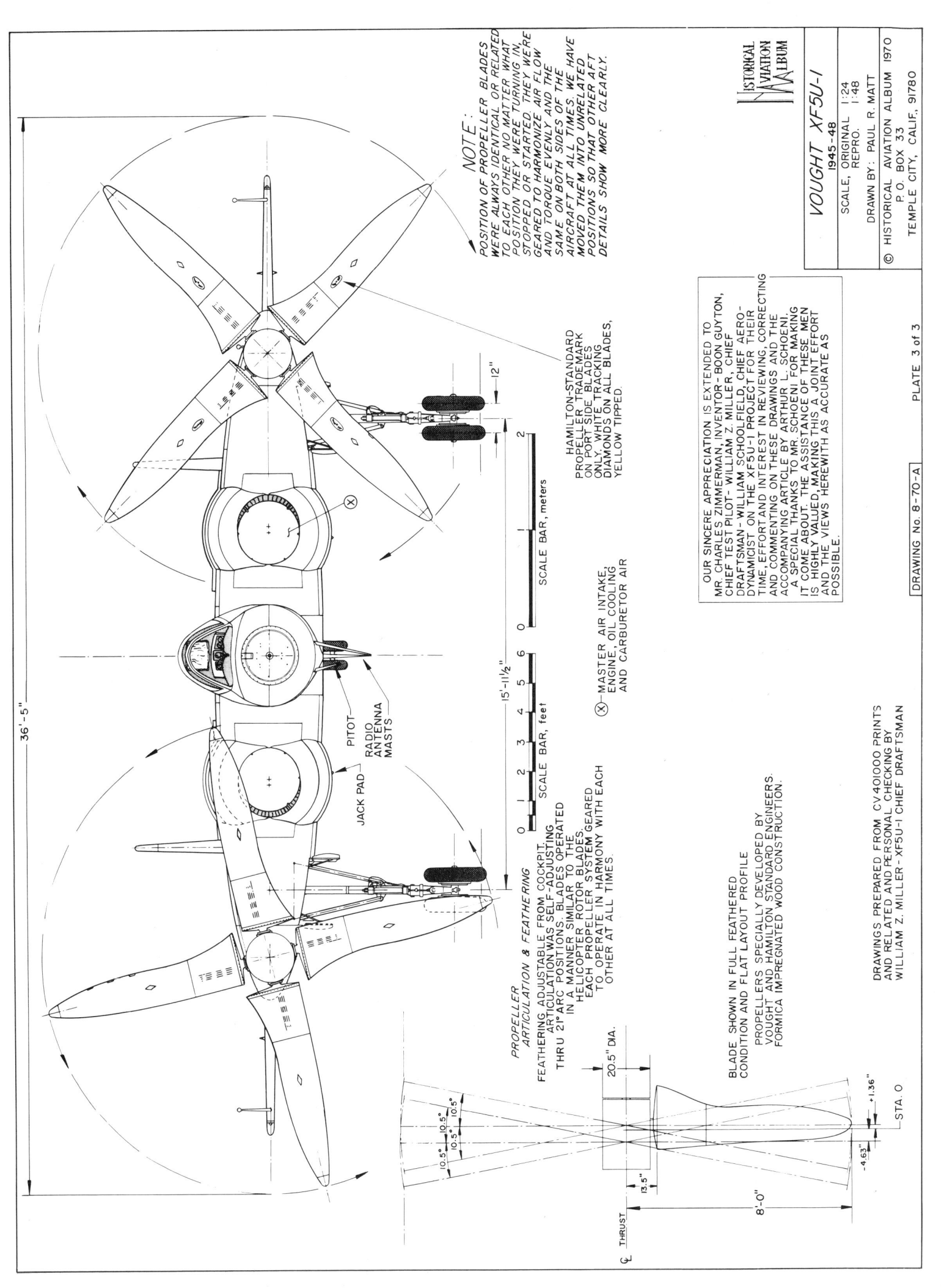

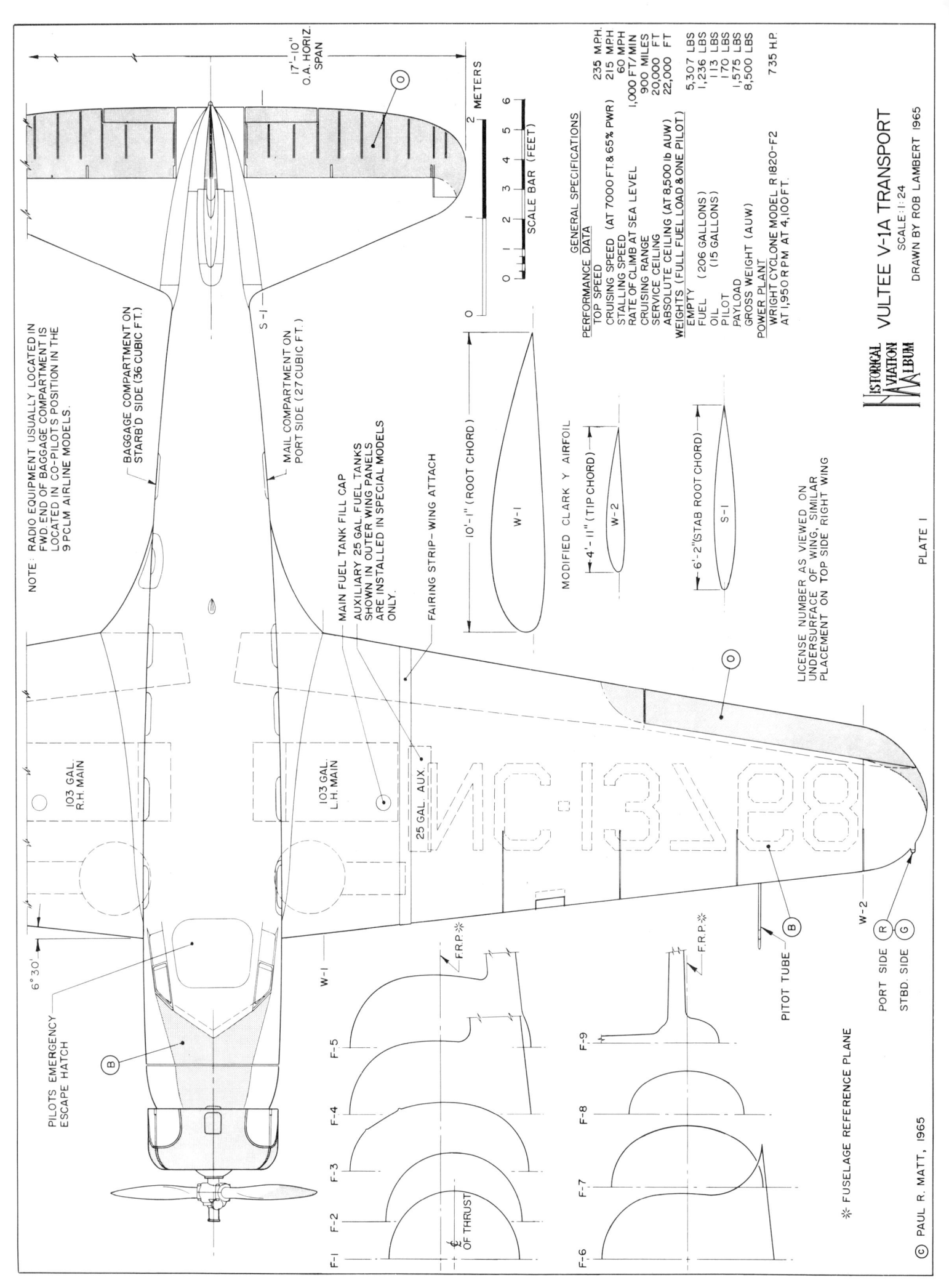

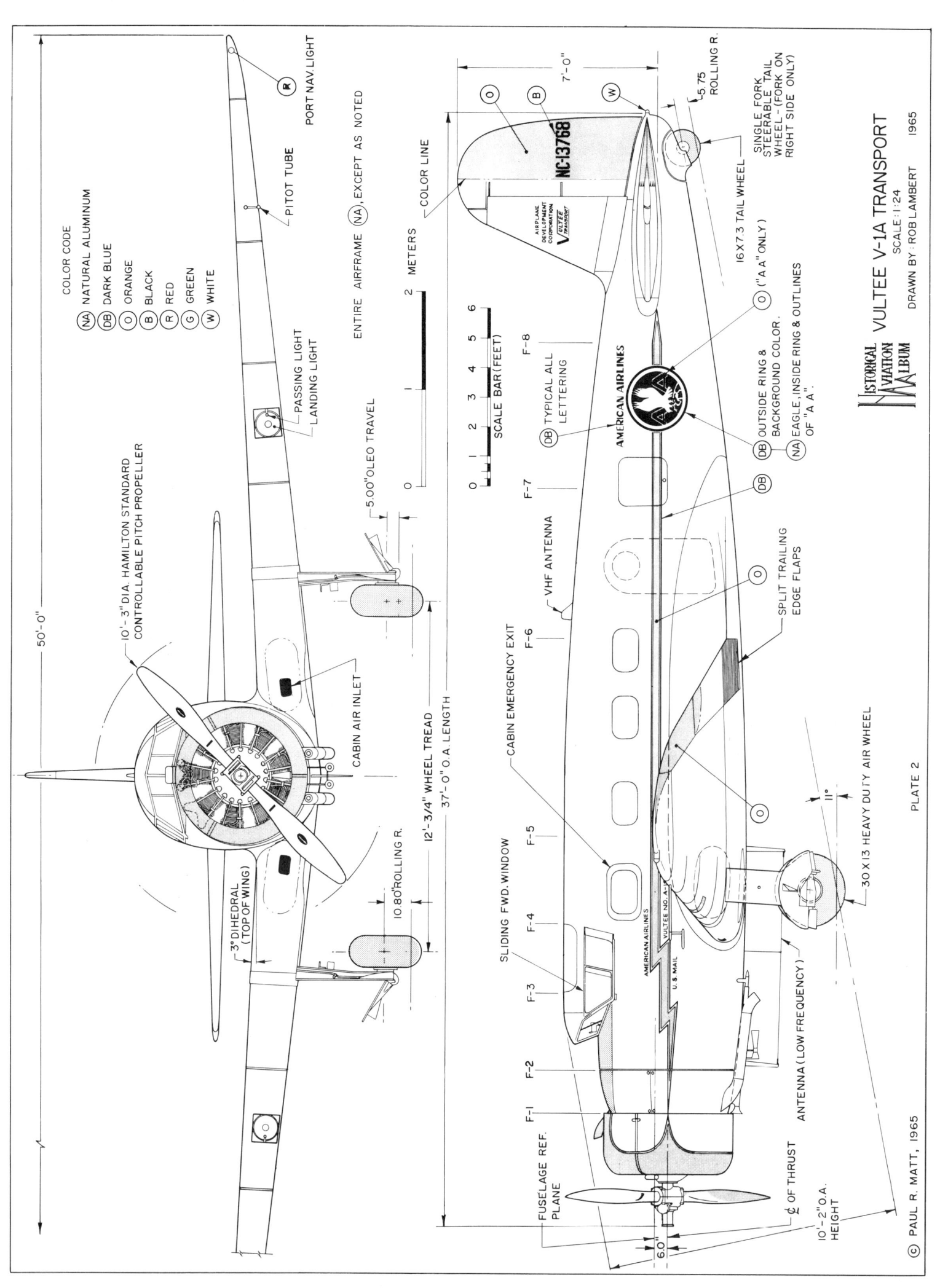

Waco YMF-5

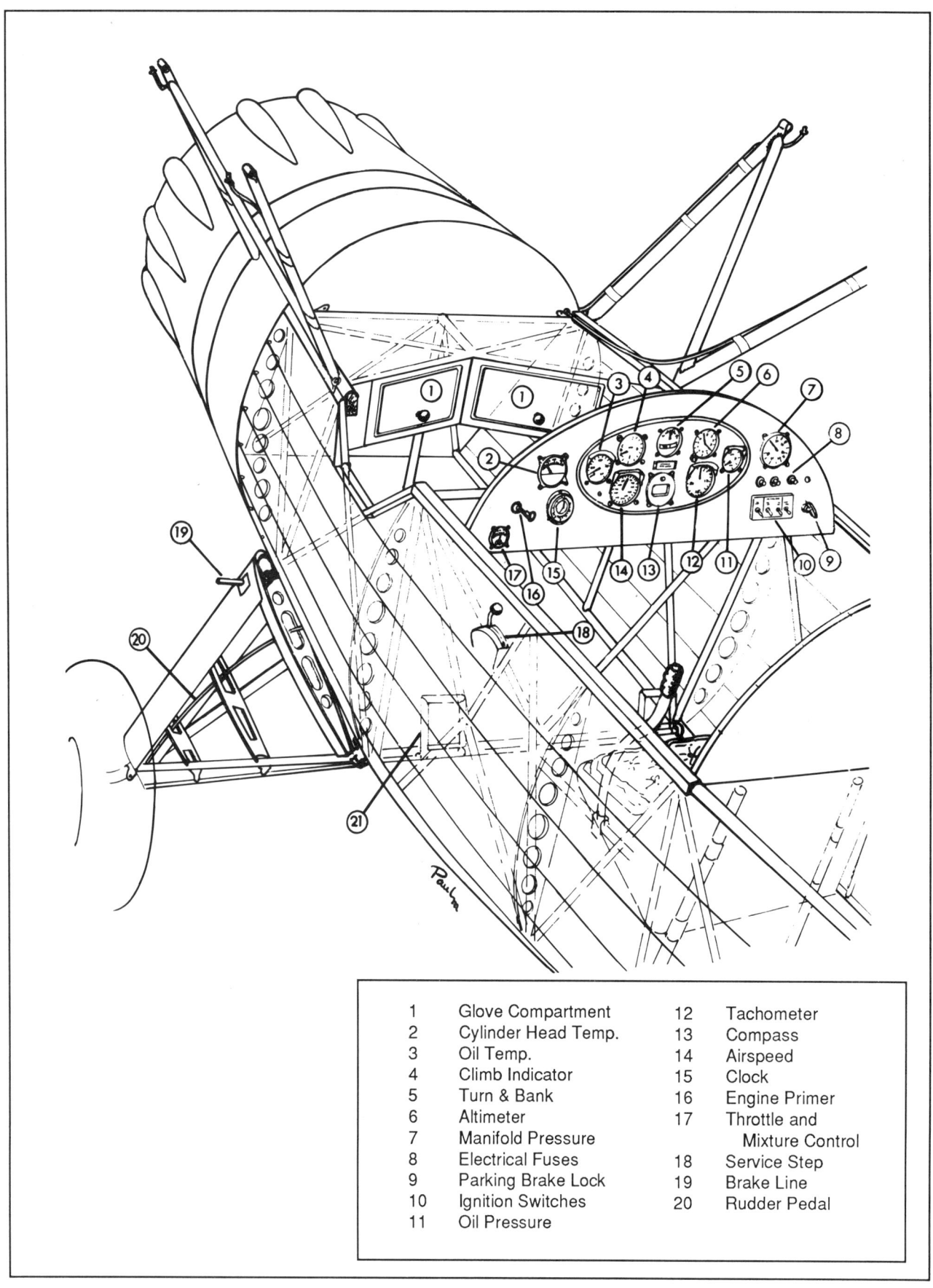

1	Glove Compartment	12	Tachometer
2	Cylinder Head Temp.	13	Compass
3	Oil Temp.	14	Airspeed
4	Climb Indicator	15	Clock
5	Turn & Bank	16	Engine Primer
6	Altimeter	17	Throttle and Mixture Control
7	Manifold Pressure	18	Service Step
8	Electrical Fuses	19	Brake Line
9	Parking Brake Lock	20	Rudder Pedal
10	Ignition Switches		
11	Oil Pressure		

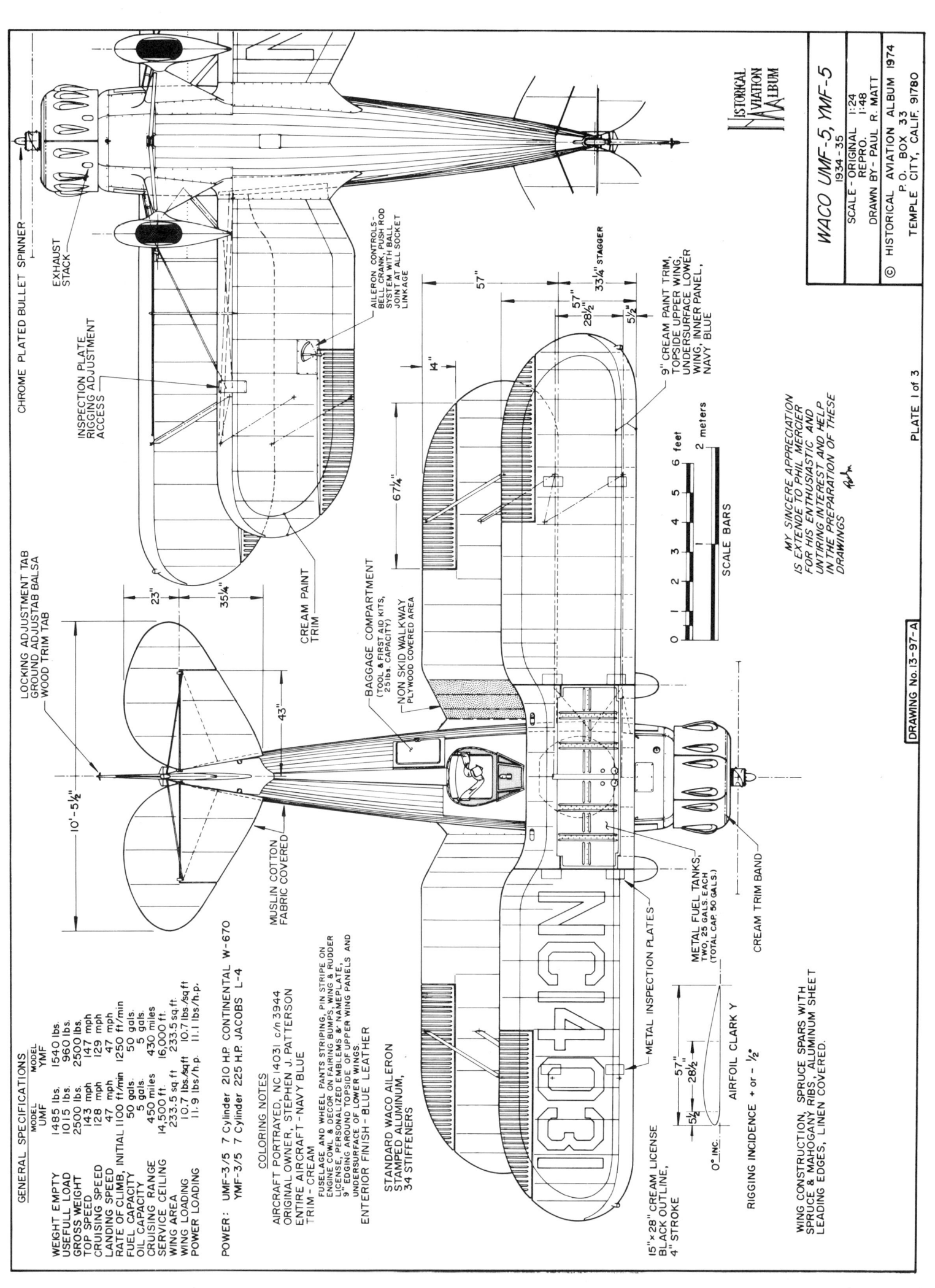

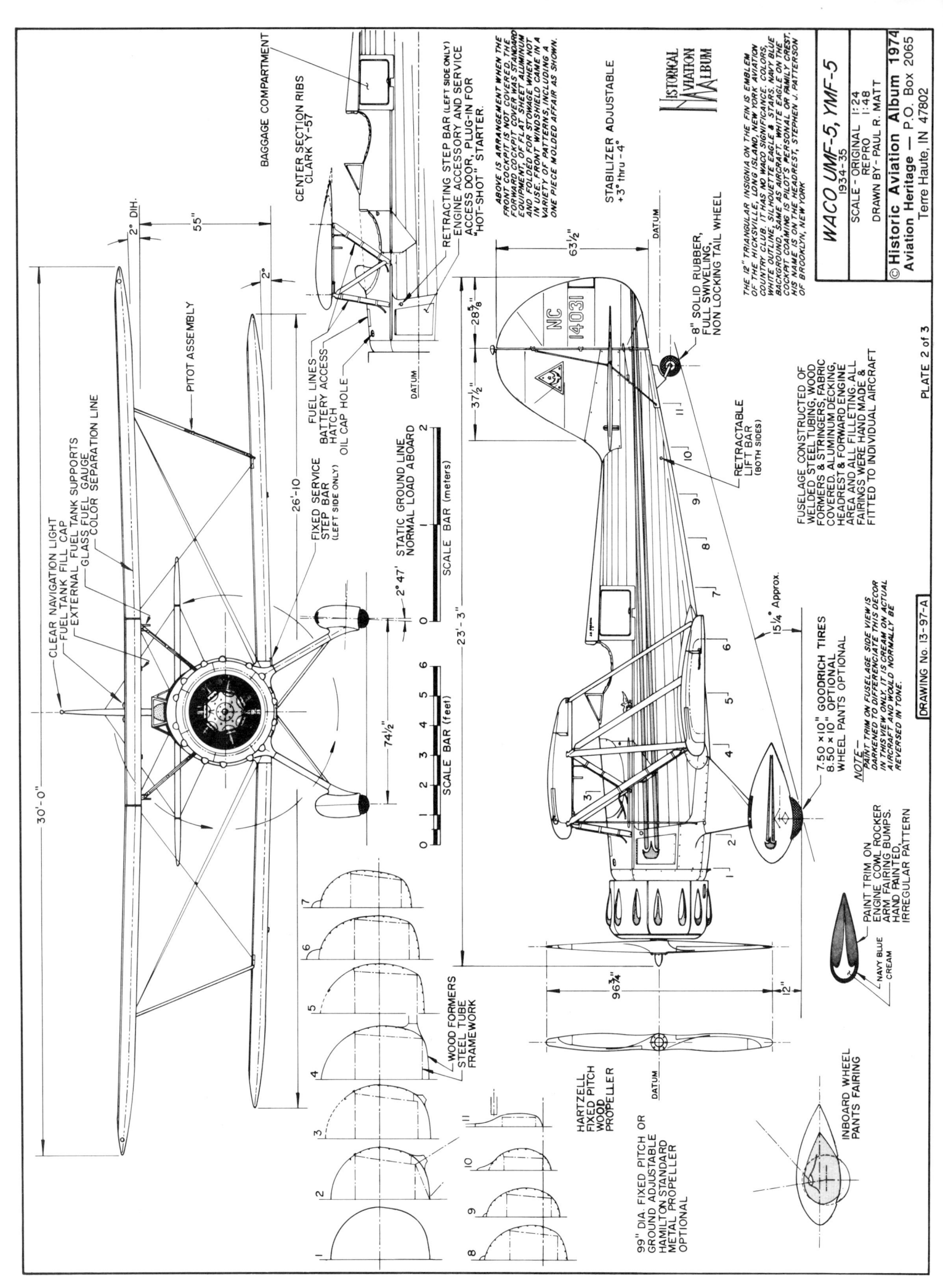

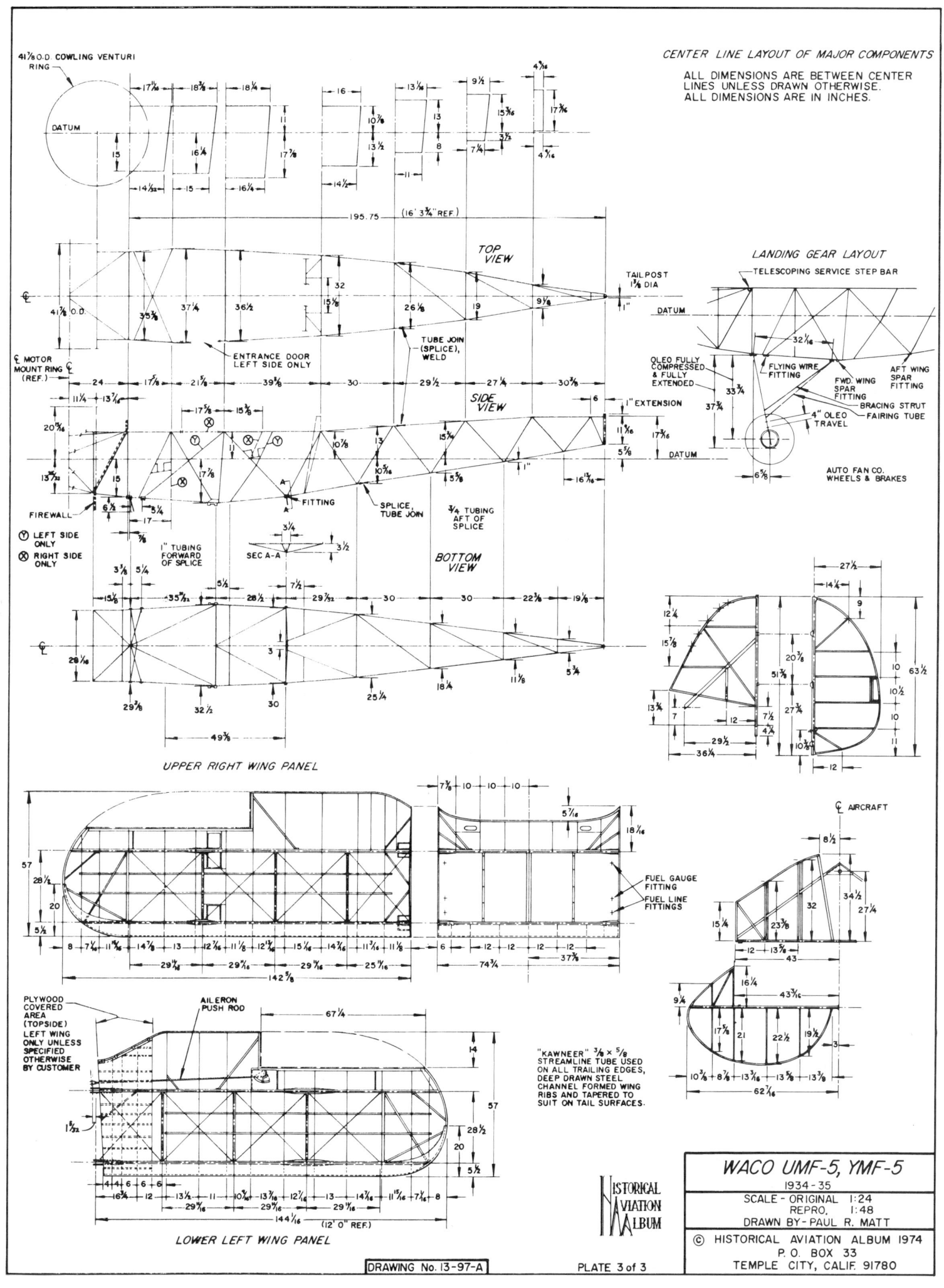

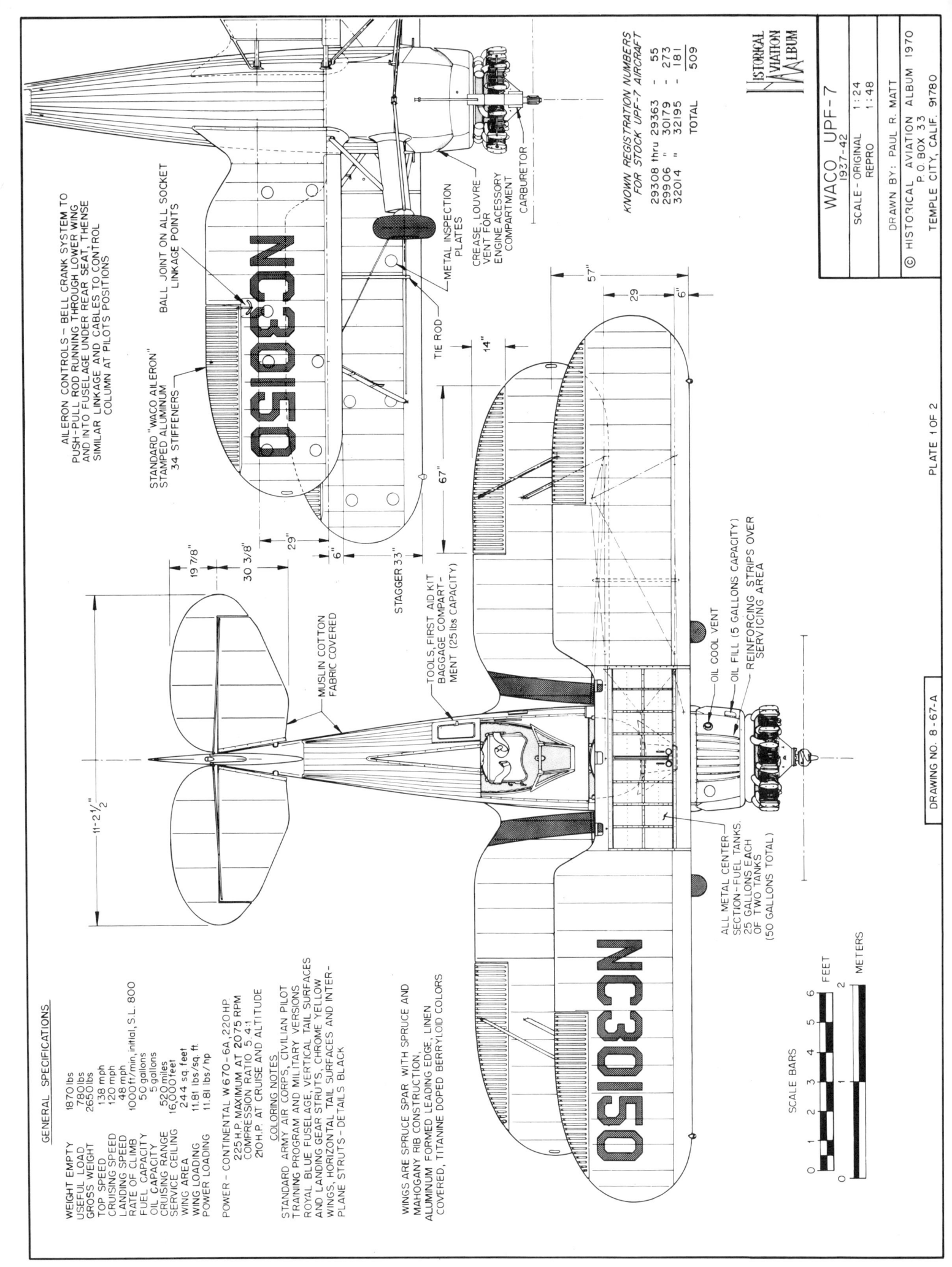

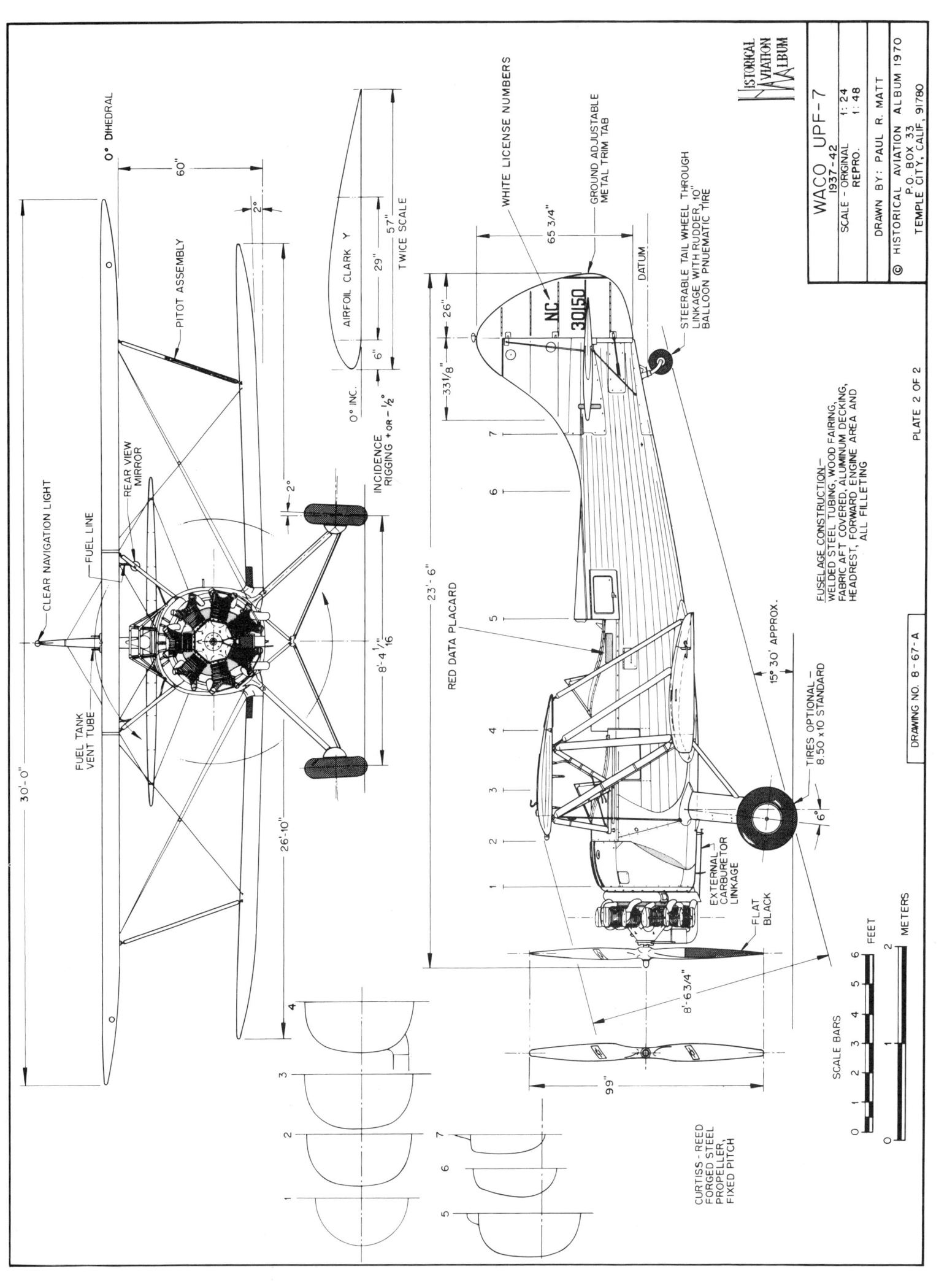

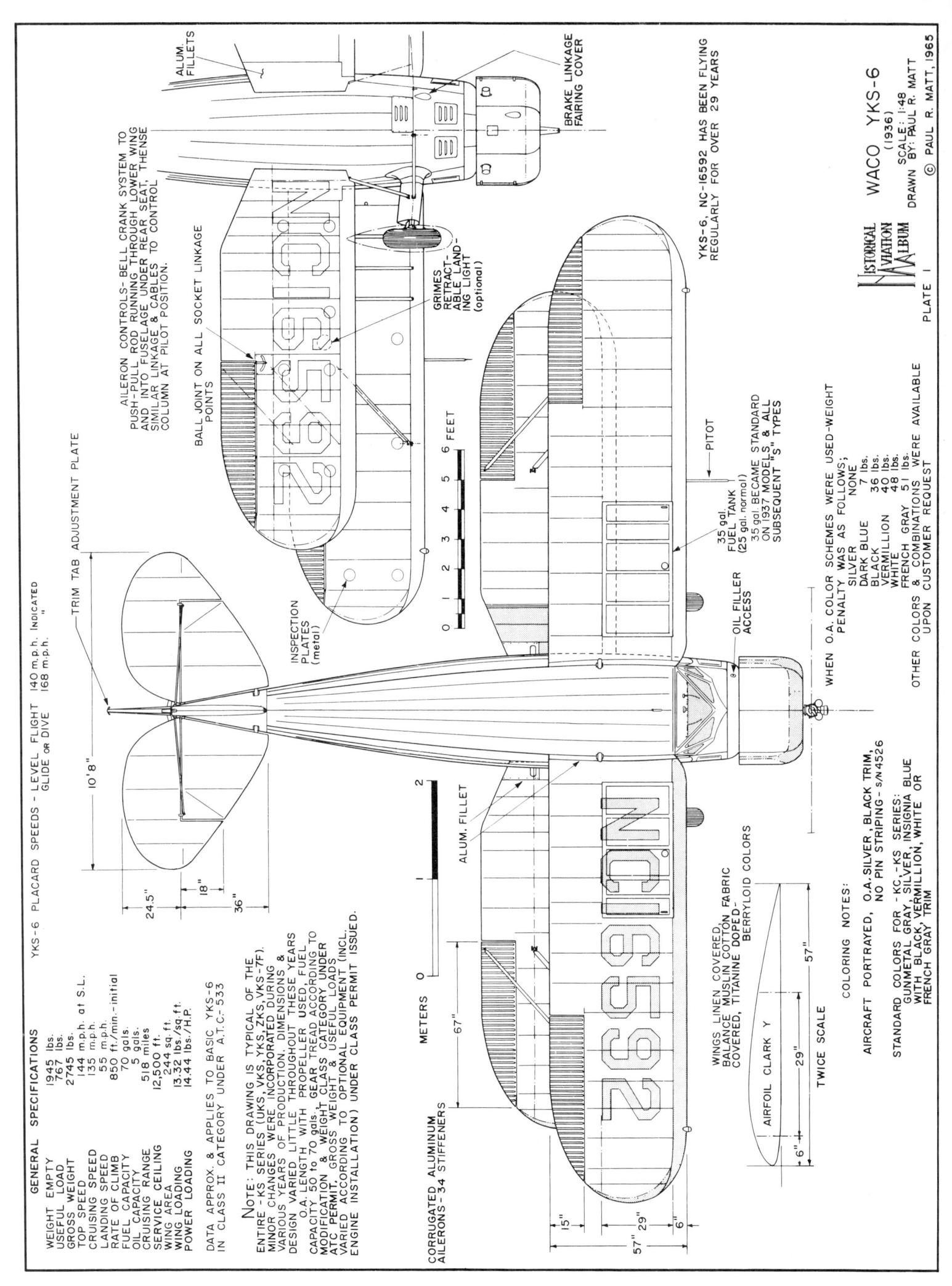

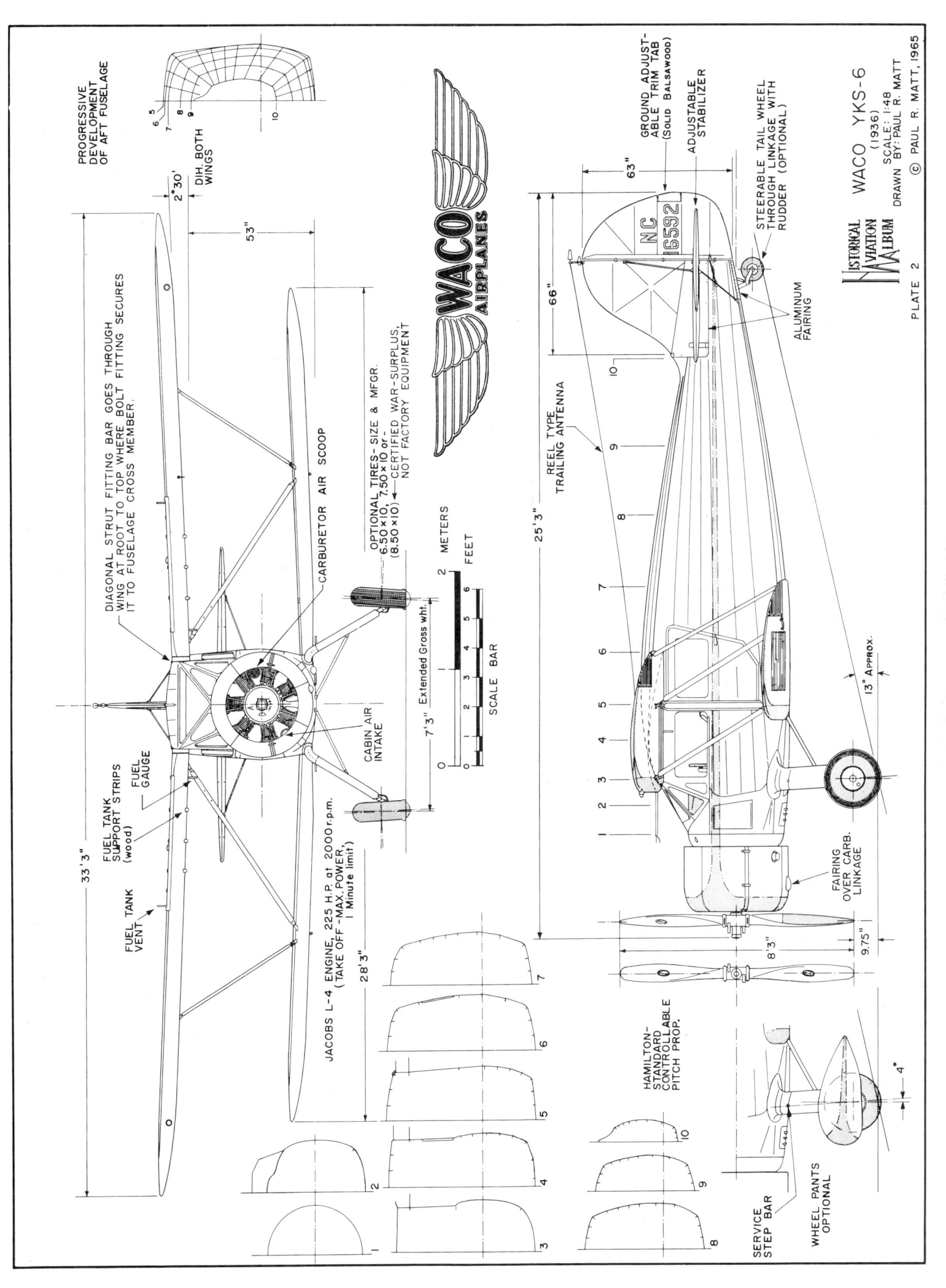

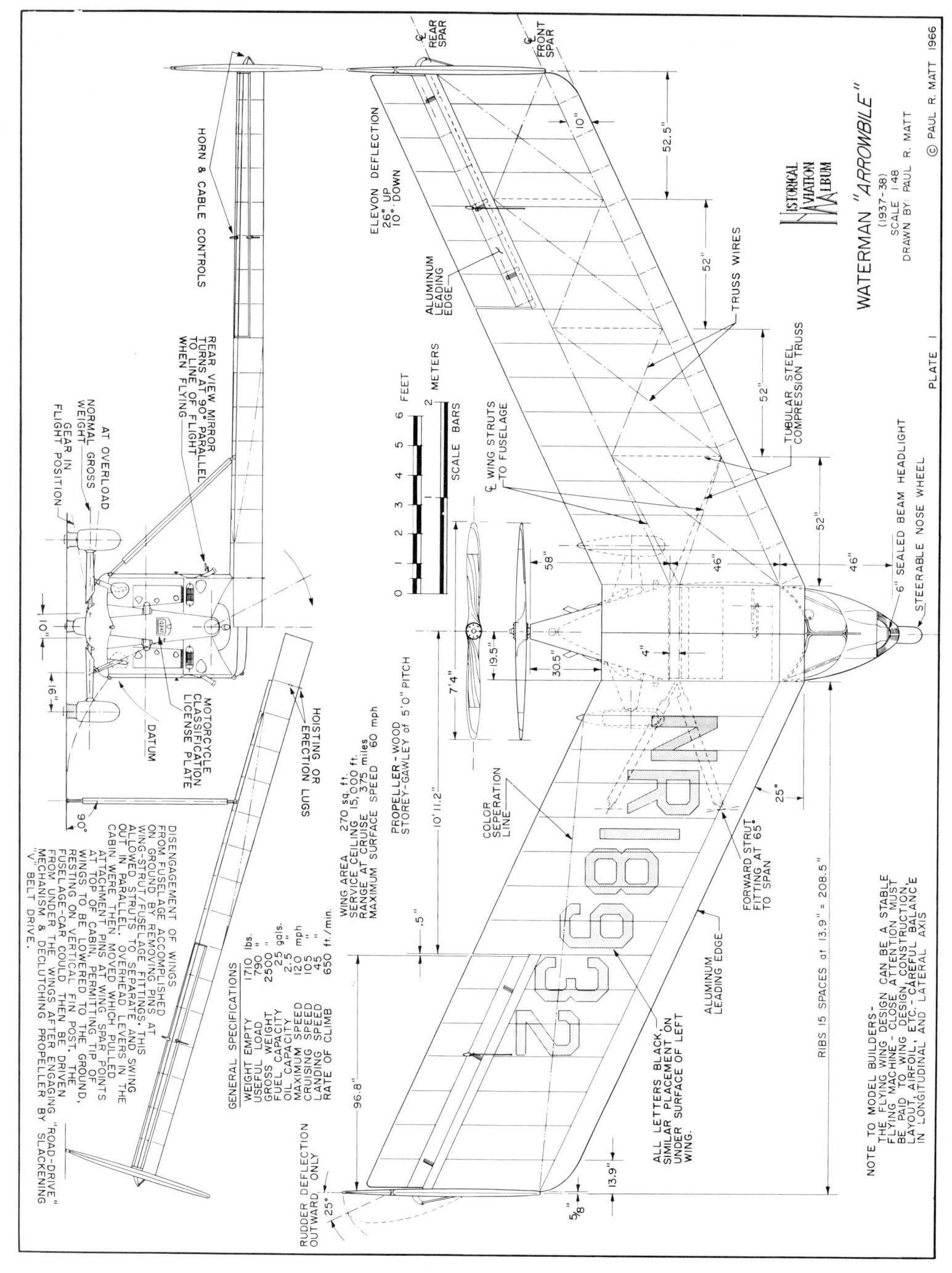

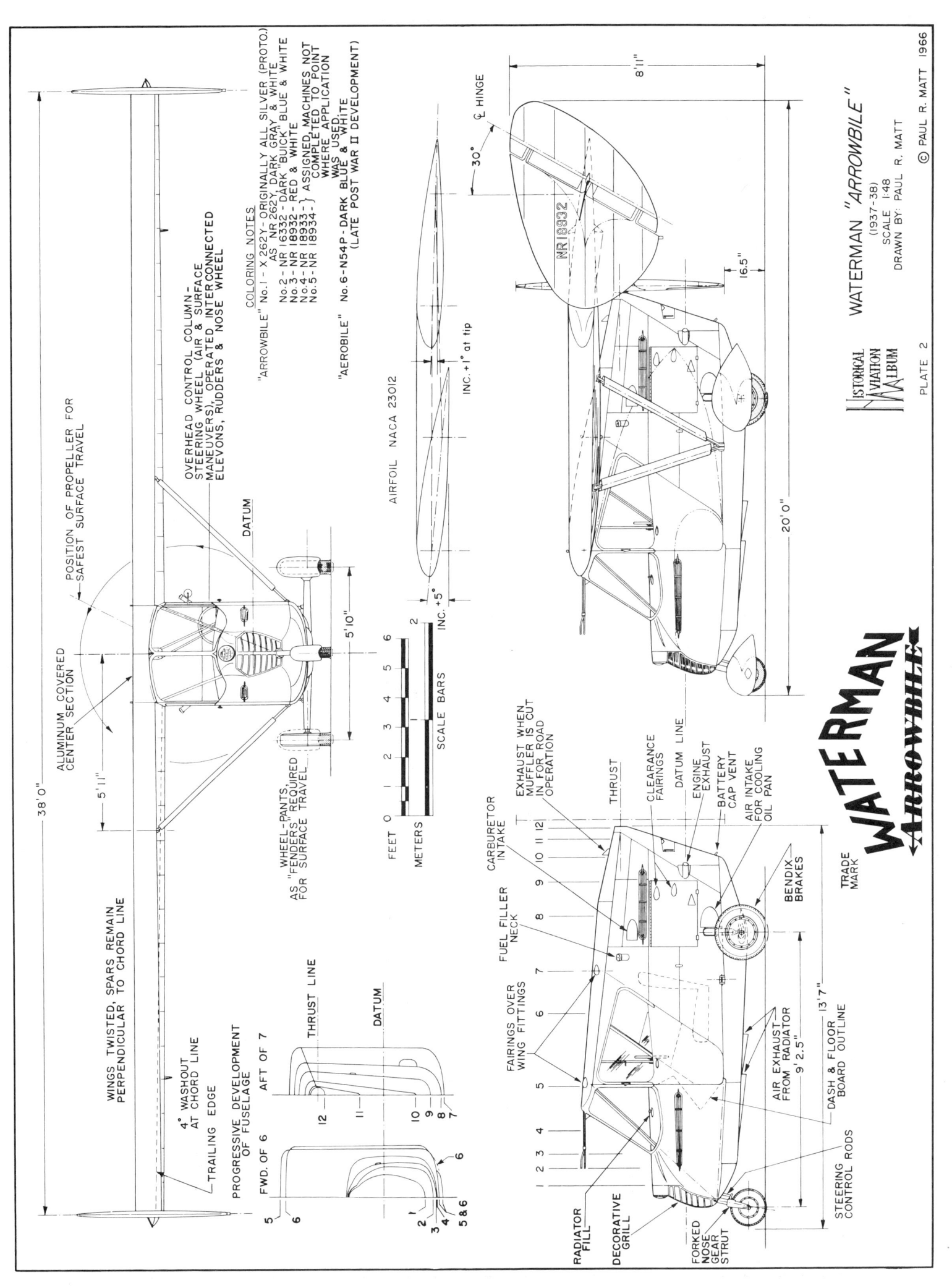

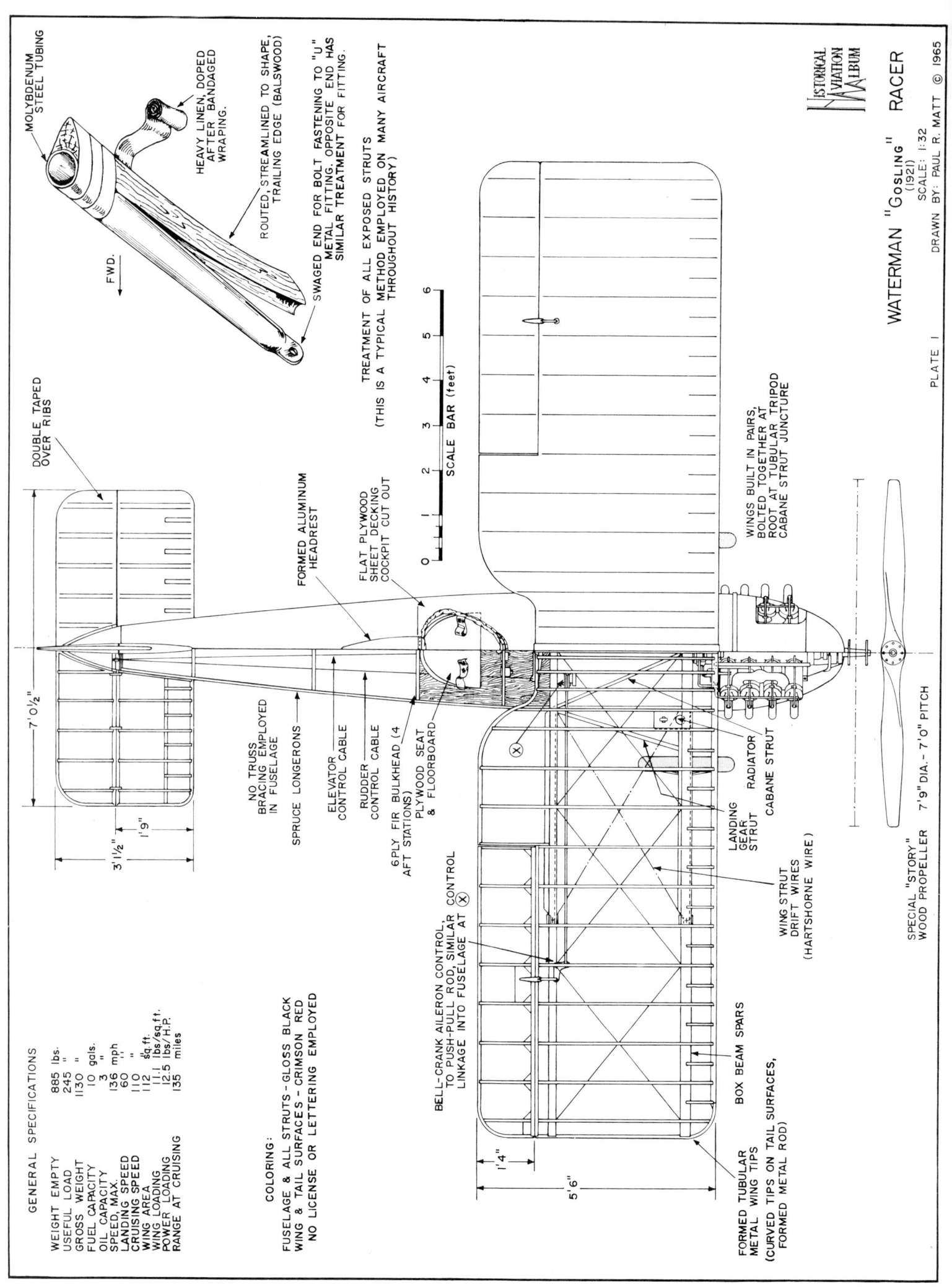

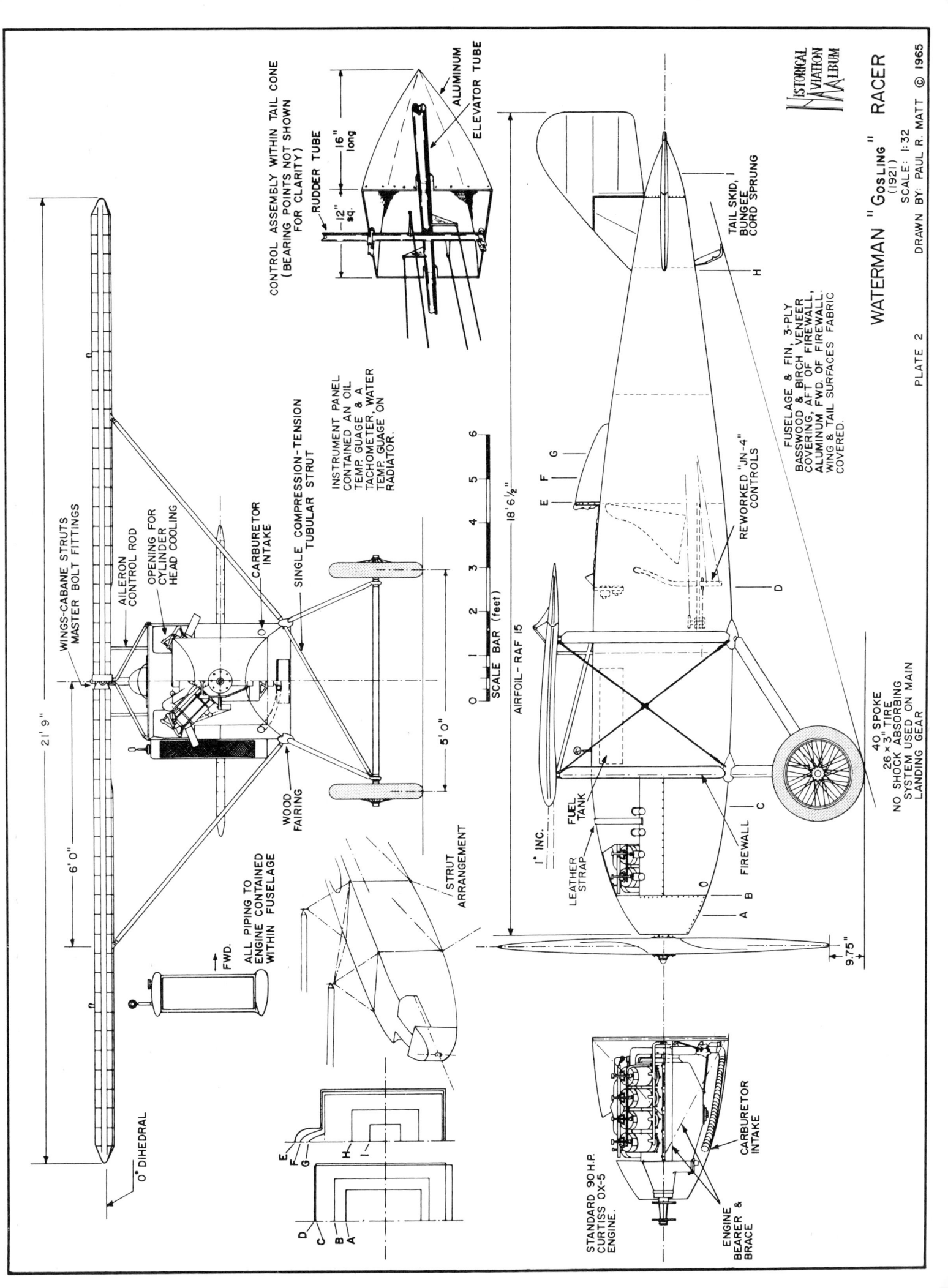

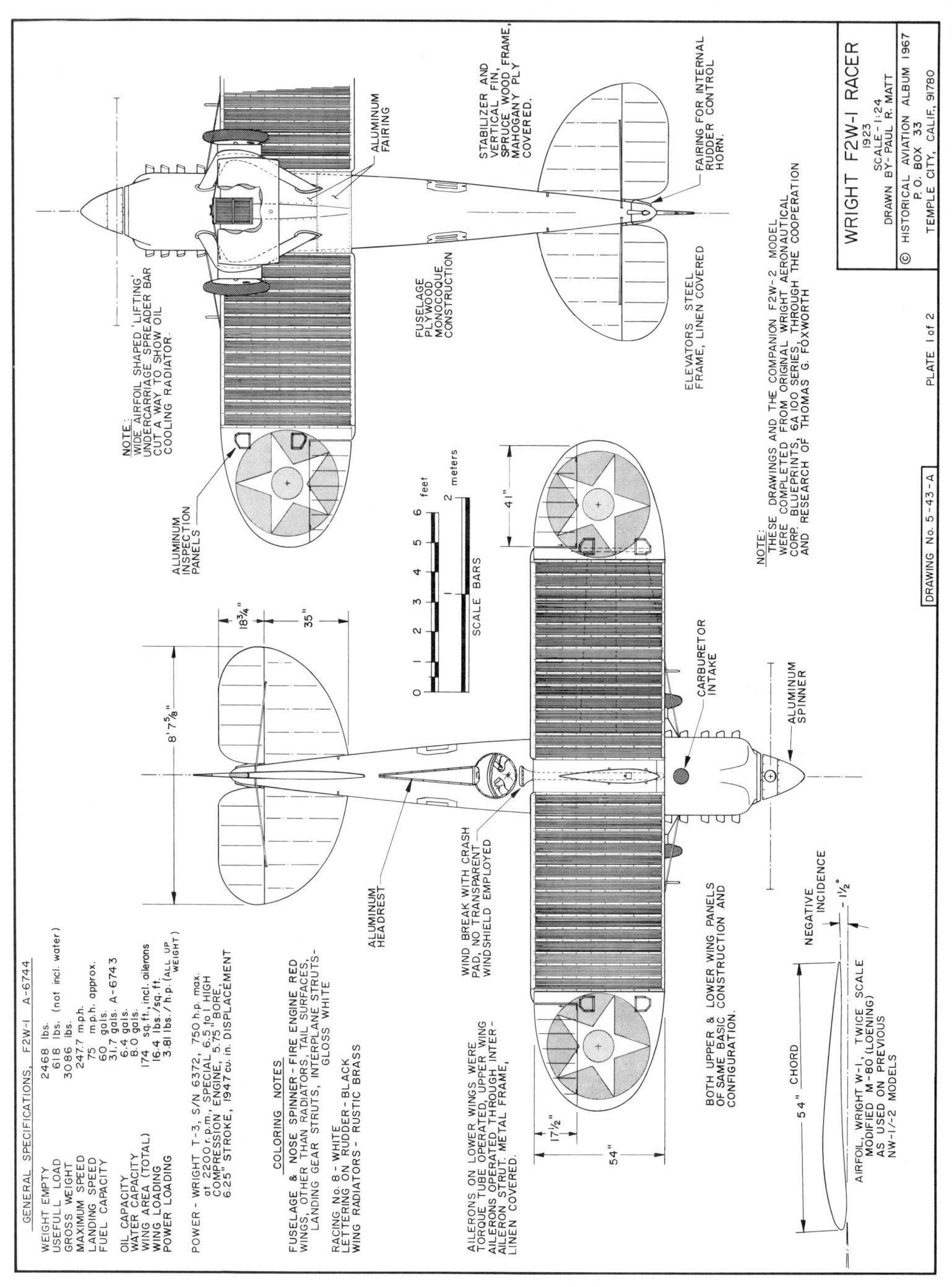

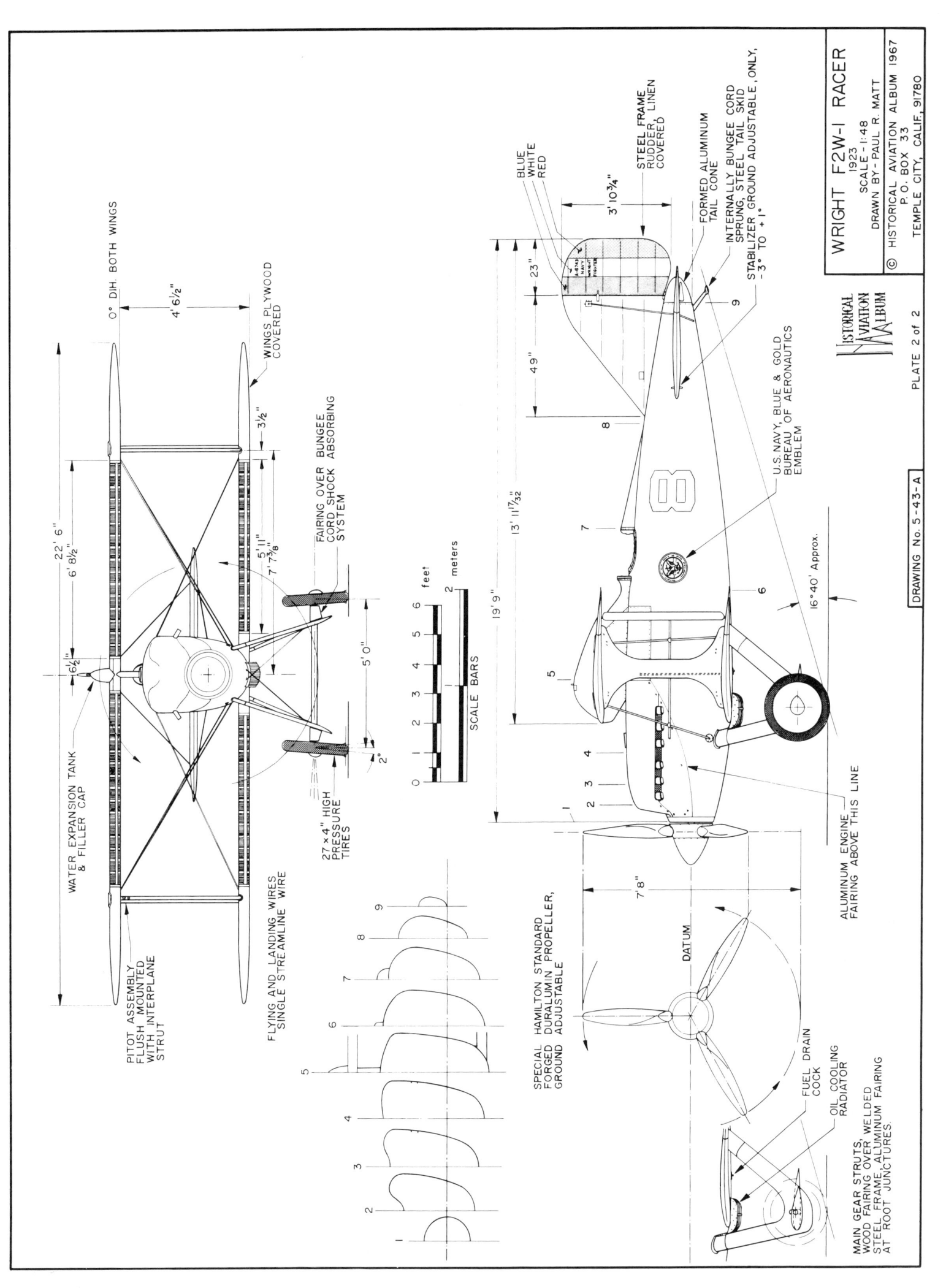

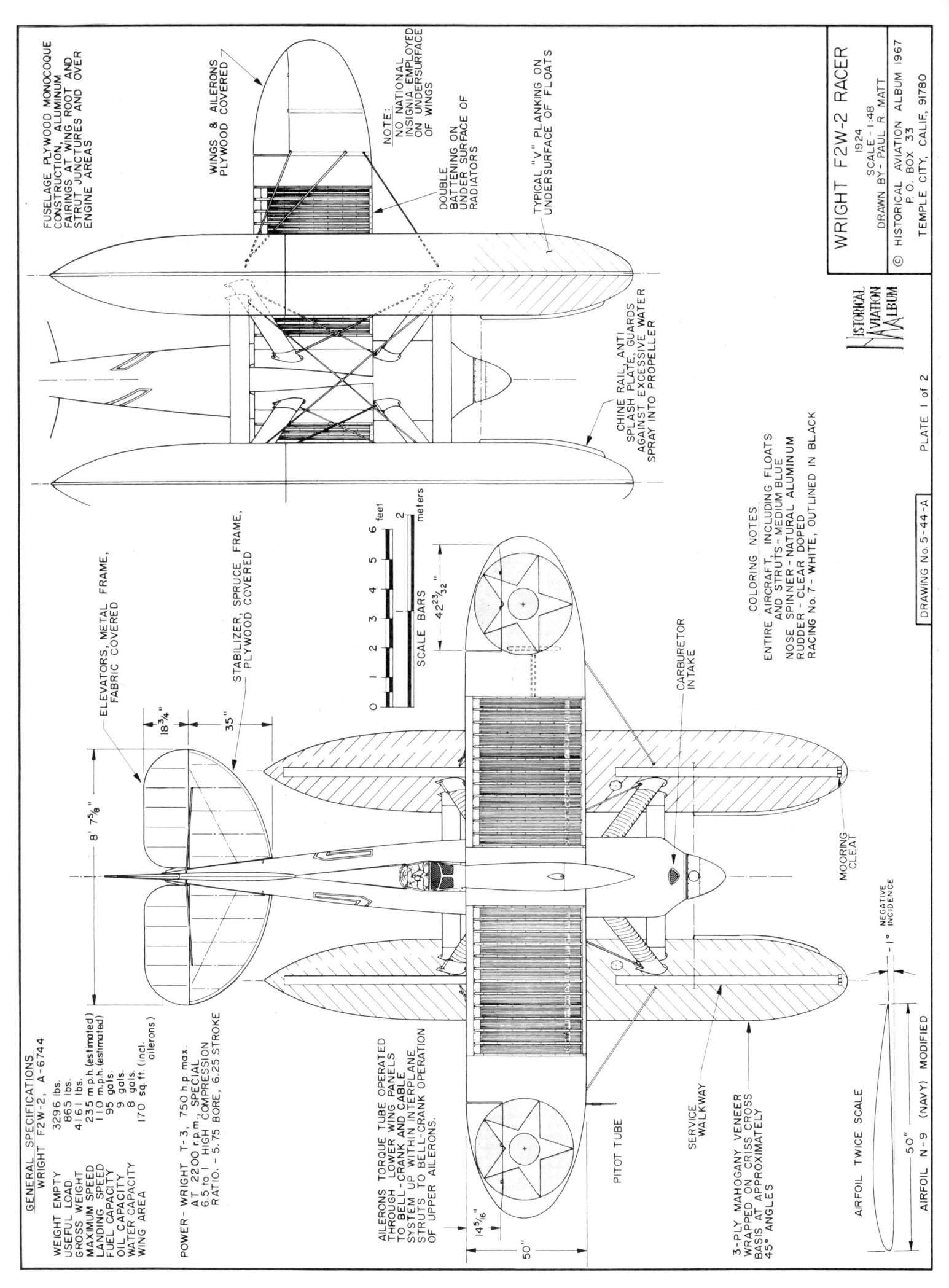

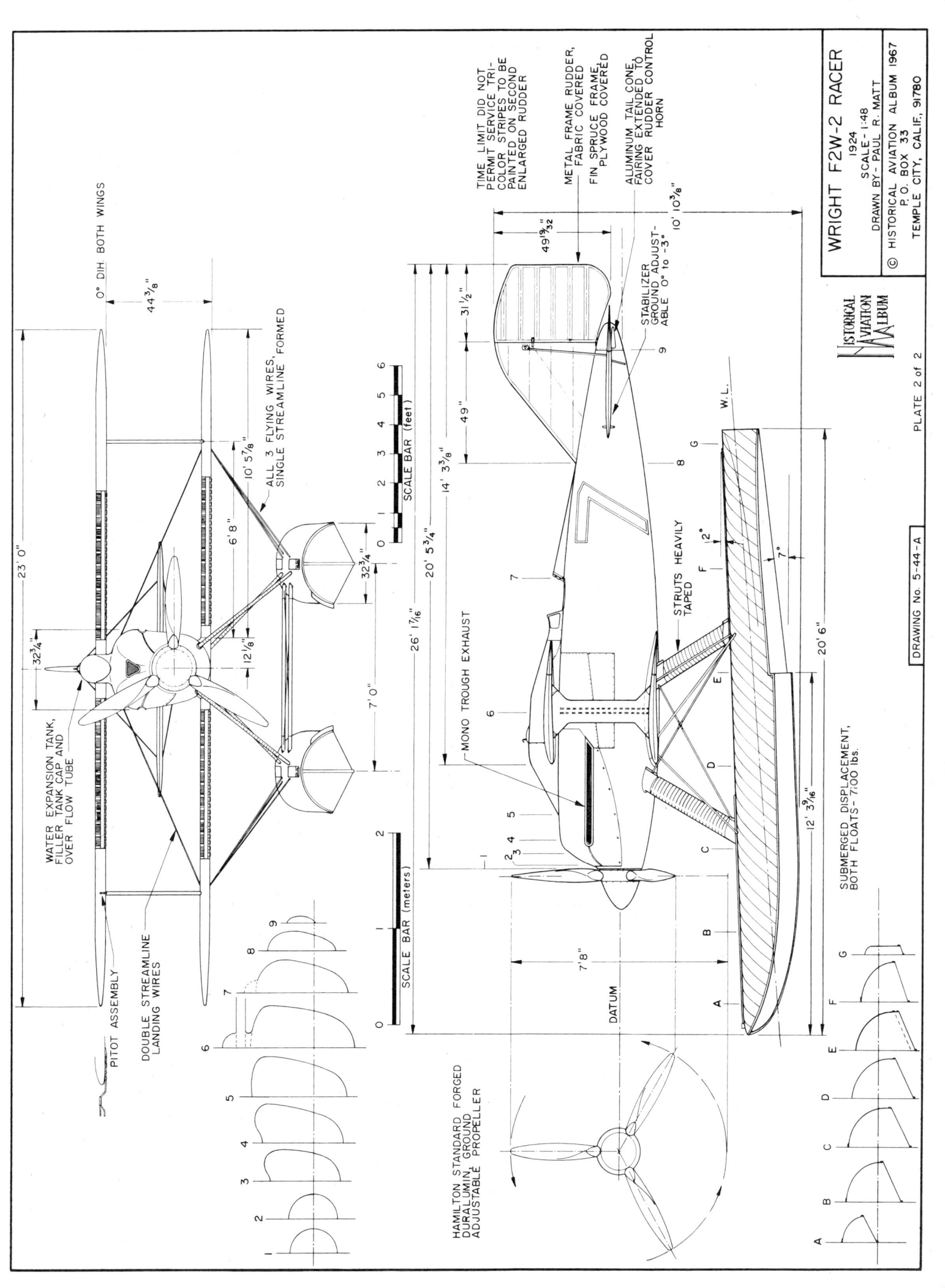

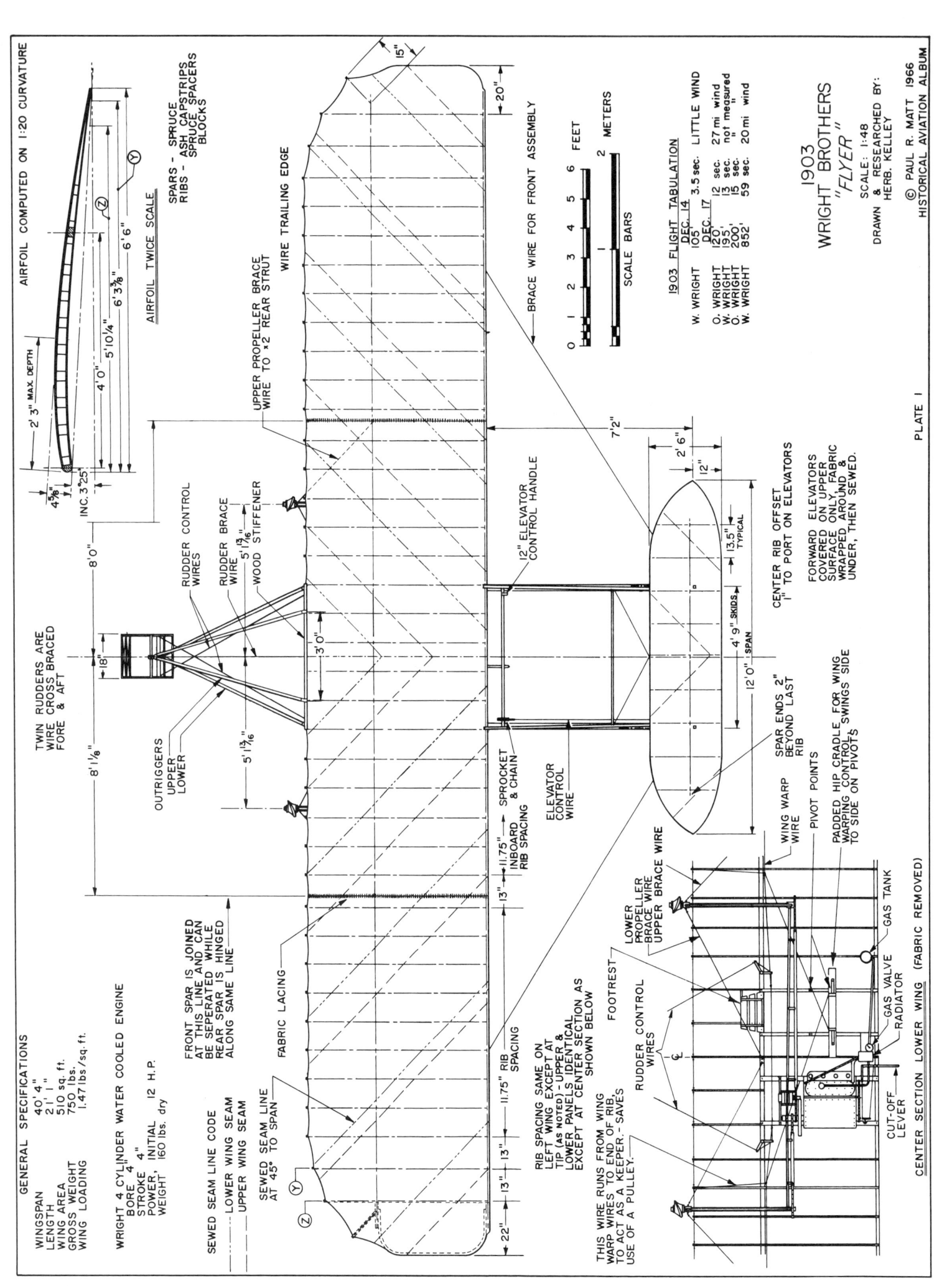

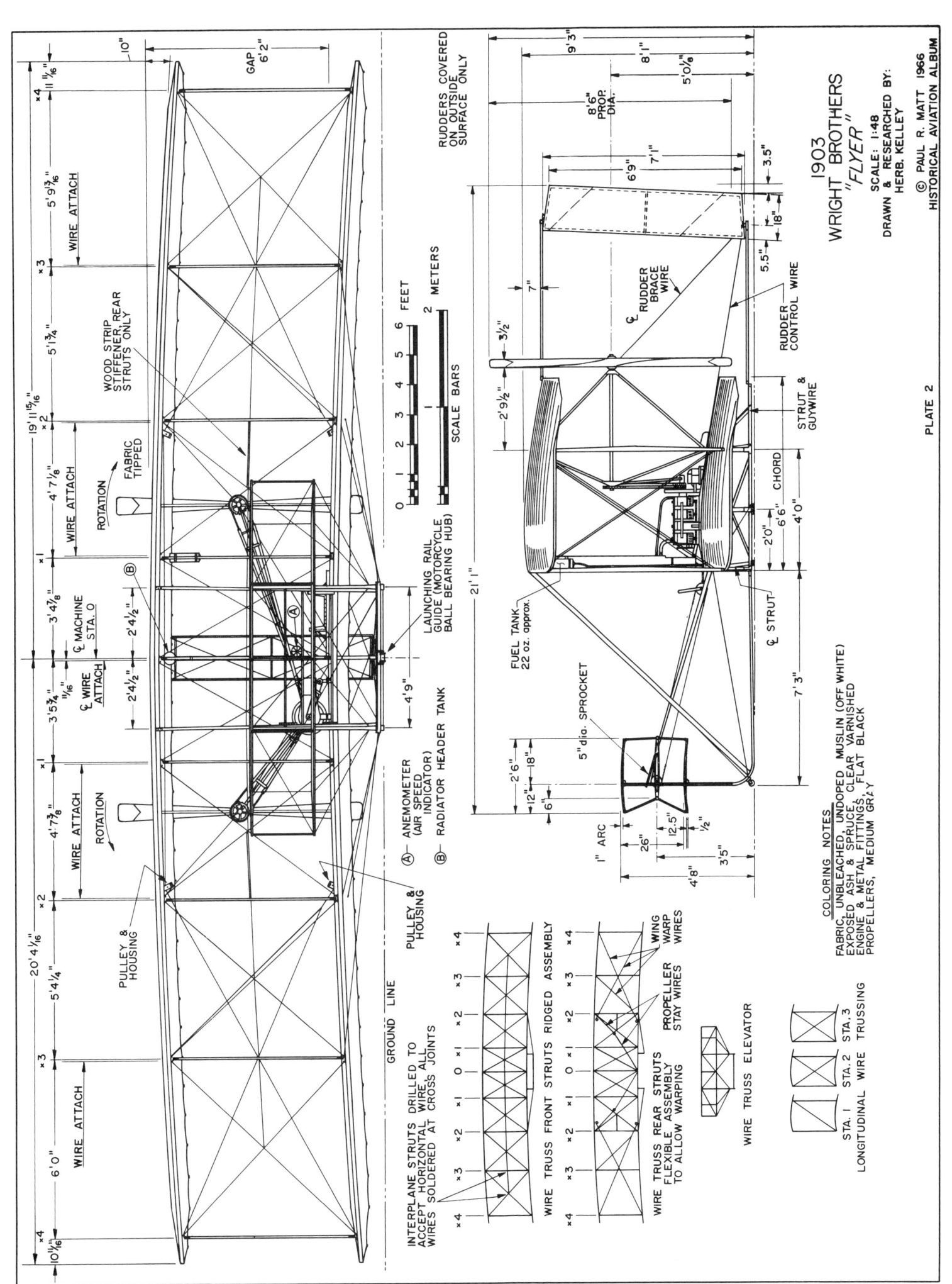

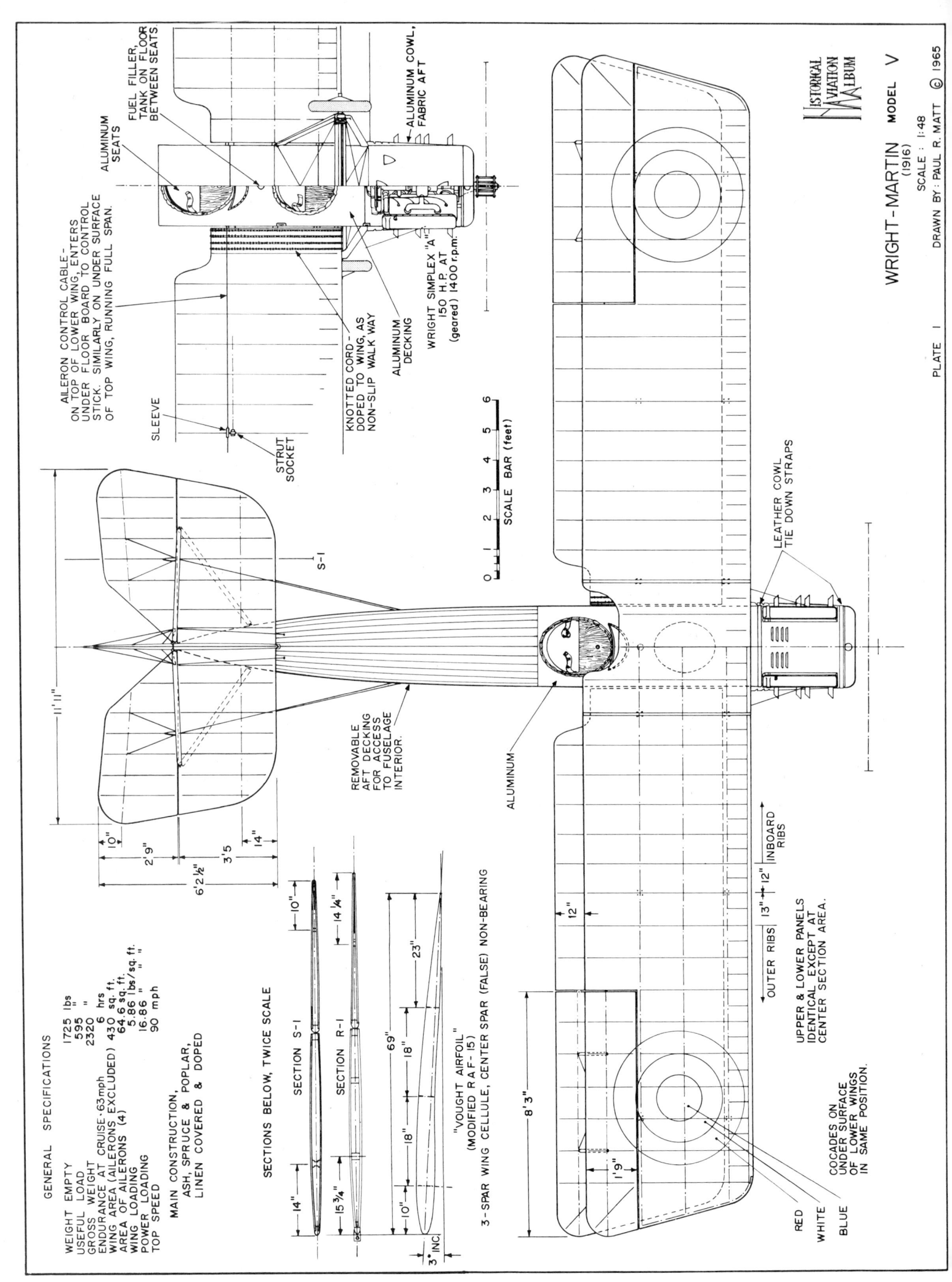

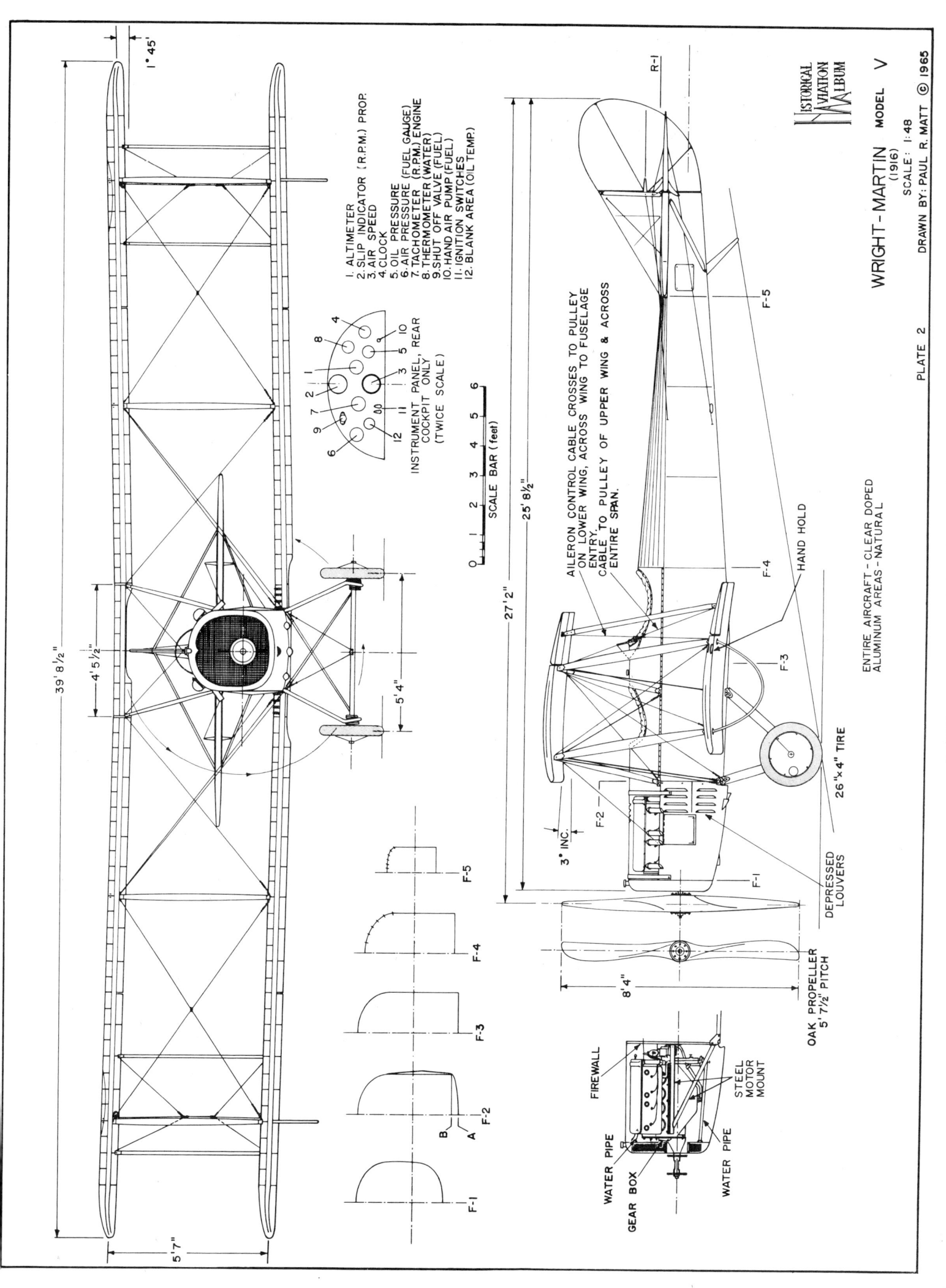

3-View Scale Drawings by Paul Matt

Volume 1 contains all airplanes listed from A through G. Volume 2 contains all airplanes listed from H through W. If you would like individual, large sheets of any of the below listed scale drawings, send request, catalog #, appropriate fee, and shipping charge of $2.50. If you want your drawing rolled in a tube – add $4.00. Clear, concise, uncluttered 3- and 4-view engineering presentations, in large (17" x 22"), easy to work with standard scales. Drawing is highly detailed and is an accurate working print with all dimensions, specifications, airfoils, cross sections, templates, and color scheme included. These drawings are part of the collection of the internationally acclaimed and highly honored *Historical Aviation Album*. All orders are folded unless a tube is ordered. Each has 2 to 6 large sheets. **Rolled in tube– add $4.00.**

DRAWING SCALE:
CODE: Last letter of catalog numer indicates scale of drawing.
A = 1:24 .... 1/2" = 1'
B = 1:16 .... 3/4" = 1'
C = 1:48 .... 1/4" = 1'
D = 1:32 .... 3/8" = 1'
E = 1:8 .... 1 1/2" = 1'
F = 1:6 .... 2" = 1'

Volume 1 contents:

_Aeromarine 39B	(#9-78A, 3 shts.)	$9.50
_Aeronca 7AC "Champion"	(#AM-41A, 2 shts.)	8.00
_Aeronca C-2	(#10-80A, 2 shts.)	8.00
_Aeronca C-3 "Collegian"	(#10-81A, 2 shts.)	8.00
_Aeronca C-3 "Master"	(#10-82A, 2 shts.)	8.00
_Aeronca K	(#15-103A, 2 shts.)	8.00
_Aeronca LB "Low wing"	(#AM-60A, 2 shts.)	8.00
_Alcor C.6.1 Jr. Transport	(#4-33D, 2 shts.)	8.00
_Anderson-Greenwood AG-14	(#12-94A, 2 shts.)	8.00
_Beechcraft D-18S Twin	(#6-59A, 3 shts.)	9.50
_Bell P-39Q "Airacobra"	(#1-7A, 2 shts.)	8.00
_Berckmans "Speed Scout"	(#2-12B, 2 shts.)	8.00
_Berliner-Joyce OJ-2	(#AM-48A, 2 shts.)	8.00
_Berliner-Joyce XF3J-1	(#15-104A, 2 shts.)	8.00
_Boeing 307 "Stratoliner"	(#7-63C, 2 shts.)	8.00
_Boeing F3B-1	(#AM-56A, 2 shts.)	8.00
_Boeing XF7B-1	(#AM-38A, 2 shts.)	8.00
_Brewster F2A-3 "Buffalo"	(#5-47A, 2 shts.)	8.00
_Cessna 120/140	(#1-9A, 2 shts.)	8.00
_Cessna C-37 "Airmaster"	(#6-50A, 2 shts.)	8.00
_Cessna T-50 "Bobcat"	(#17-115A, 2 shts.)	8.00
_Consolidated P2Y-2	(#9-71C, 3 shts.)	9.50
_Consolidated PBY-5A "Catalina"	(#17-114C, 4shts.)	11.00
_Curtiss "Carrier Pigeon I"	(#2-15A, 2 shts.)	8.00
_Curtiss 1st Milit. Tract. S.C. No. 21/22 1913.	(#6-55A, 2 shts.)	8.00
_Curtiss A-3B "Falcon"	(#AM-39A, 2 shts.)	8.00
_Curtiss AT-9 "Jeep"	(#2-18A, 2 shts.)	8.00
_Curtiss B-2 Condor Bomber	(#18-116C, 2 shts.)	8.00
_Curtiss B-20 Condor Transport	(#18-117C, 2 shts.)	8.00
_Curtiss F Boat	(#1-1A, 2 shts.)	8.00
_Curtiss F92 "Sparrowhawk"	(#18-118A, 2 shts.)	8.00
_Curtiss MF-K-6 "Seagull"	(#AM-37A, 2 shts.)	8.00
_Curtiss P-36 "Hawk"	(#7-62A, 2 shts.)	8.00
_Curtiss P-6E "Hawk"	(#5-42A, 2 shts.)	8.00
_Curtiss PW-8	(#10-79A, 2 shts.)	8.00
_Curtiss R-6 Racer	(#HS-73A, 2 shts.)	8.00
_Curtiss SC-1 "Seahawk"	(#1-8A, 2 shts.)	8.00
_Curtiss SNC-1 "Falcon"	(#AA-65A, 2 shts)	8.00
_Curtiss SO3C-1 "Seagull"	(#AM-34A, 2 shts.)	8.00
_Curtiss Twin JN	(#3-21A, 2 shts.)	8.00
_Curtiss-Cox "Texas Wildcat"	(#2-14B, 2 shts.)	8.00
_Curtiss-Wright "Condor II"	(#1-6C, 2 shts)	8.00
_Curtiss-Wright CW-1 "Jr."	(#11-88A, 3 shts.)	9.50
_Douglas 0-2	(#11-86A, 3 shts.)	9.50
_Douglas 0-2H	(#11-87A, 2 shts.)	8.00
_Douglas 0-38	(#12-92A, 2 shts.)	8.00
_Douglas 0-38E	(#12-93A, 2 shts.)	8.00
_Douglas M-2	(#14-68A, 2 shts.)	8.00
_Douglas A-20G "Havoc"	(#15-102D, 6 shts.)	13.00
_Etrich 1913 Taube	(#TB-105D, 3 shts.)	9.50
_Fairchild FC-1	(#17-111A, 2 shts.)	8.00
_Fairchild FC-2	(#17-112A, 2 shts.)	8.00
_Fairchild FC-2W "Stars & Stripes"	(#17-113A, 2 shts.)	8.00
_Fairchild M-62, PT-19 "Cornell"	(#AA-69A, 2 shts.)	8.00
_Fokker T.5 Netherlands Bomber	(#AA-61C, 2 shts.)	8.00
_Gallaudet D-1	(#3-20A, 2 shts.)	8.00
_General Aviation Clark GA-43	(#13-95D, 2 shts.)	8.00
_Grumman F-11F-1 "Tiger"	(#2-19D, 2 shts.)	8.00
_Grumman FF-1	(#AA-77A, 2 shts.)	8.00
_Grumman G-44 "Widgeon"	(#AM5-30A, 2 shts.)	8.00
_Grumman J2F-5 "Duck"	(#6-53A, 3 shts.)	9.50

Volume 2 contents:

_Heath LNB-4 Parasol	(#4-35B, 2 shts.)	$8.00
_Howard DGA-15P	(#AM-52A, 2 shts.)	8.00
_Howard DGA-3 "Pete"	(#12-89A, 2 shts.)	8.00
_Howard DGA-4 "Mike"	(#13-90A, 2 shts.)	8.00
_Howard DGA-5 "Ike"	(#13-91A, 2 shts.)	8.00
_Howard DGA-6 "Mr. Mulligan"	(#14-99A, 3 shts.)	9.50
_Hughes 1B Longwing Racer	(#16-108A, 3 shts.)	9.50
_Hughes 1B Shortwing Racer	(#16-107A, 3 shts.)	9.50
_Laird LC-DW-300 "Solution"	(#7-64C, 2 shts.)	8.00
_Laird LC-DW-500 "Super Solution"	(#8-74A, 2 shts.)	8.00
_Laird-Turner LTR-14	(#10-83A, 2 shts.)	8.00
_Lavochkin LA-7	(#AA-46A, 4 shts.)	11.00
_Lockheed F-80 "Shooting Star"	(#2-10A, 2 shts.)	8.00
_Lockheed Model 9 "Orion"	(#3-22A, 2 shts.)	8.00
_Lockheed PV-1 "Ventura"	(#E-101C, 3 shts.)	9.50
_LWF Cato Model L "Butterfly"	(#10-84A, 2 shts.)	8.00
_LWF Model G-2	(#2-13A, 2 shts.)	8.00
_LWF Model H "Owl"	(#11-85C, 2 shts.)	8.00
_Martin BM-1/2	(#1-5A, 2 shts.)	8.00
_Martin T4M-1, Great Lakes TG-1	(#4-29D, 2 shts.)	8.00
_Martin TT, 1913 Trainer	(#5-40A, 2 shts.)	8.00
_Messerschmitt ME 109E-3	(#AA-75A, 2 shts.)	8.00
_Morehouse 2 Cyl. Aero Engine	(#3-27E, 1 sht.)	8.00
_Navy-Wright NW-1 Mystery Racer	(#4-31A, 2 shts.)	8.00
_Navy-Wright NW-2 Mystery Racer	(#4-32A, 2 shts.)	8.00
_North American 0-47A	(#13-98A, 3 shts.)	9.50
_North American AT-6D	(#16-106A, 3 shts.)	9.50
_North American XB-70-1 "Valkyrie"	(#7-66G, 3 shts.)	9.50
_Packard-Lepere LUSAC-11	(#1-3A, 2 shts.)	8.00
_Pfitzner 1910 Monoplane	(#2-11A, 2 shts.)	8.00
_Piper J-3 "Cub"	(#18-121A, 2 shts.)	8.00
_Piper J-4 "Cub Coupe"	(#18-122A, 2 shts.)	8.00
_Piper PA-12 "Super Cruiser"	(#4-36A, 2 shts.)	8.00
_Republic RC-3 Seabee	(#16-109A, 3 shts.)	9.50
_Rover Inverted Aero Engine	(#5-45F, 1 sht.)	5.00
_Ryan B-5 "Brougham"	(#RB-96A, 2 shts.)	8.00
_Ryan FR-1 "Fireball"	(#3-26A, 2 shts.)	8.00
_Ryan SCW Low Wing	(#AM9-49A, 2 shts.)	8.00
_Ryan ST-A	(#9-76A, 2 shts.)	8.00
_Seversky BT-8	(#3-23A, 2 shts.)	8.00
_Seversky P-35	(#AA-57A, 2 shts.)	8.00
_Sikorsky S-39B	(#14-100A, 5 shts.)	10.50
_Standard J-1	(#17-110A, 2 shts.)	8.00
_Taylor E-2 "Cub"	(#18-119A, 2 shts.)	8.00
_Taylor J-2 "Cub"	(#18-120A, 2 shts.)	8.00
_Thomas-Boeing MB-3A		
_Thomas-Morse MB-3	(#4-28B, 2 shts.)	8.00
_Timm TC-170 "Collegiate"	(#8-72A, s shts.)	8.00
_Verville R-3 Racer	(#6-58A, 2 shts.)	8.00
_Vought F4U-1 "Corsair"	(#AA-51A, 3 shts.)	9.50
_Vought SBU-1	(#3-24A, 2 shts.)	8.00
_Vought XF5U	(#8-70A, 3 shts.)	9.50
_Vultee V-1A	(#2-16A, 2 shts.)	8.00
_Waco UMF/YMF-5	(#13-97A, 3 shts.)	9.50
_Waco UPF-7	(#8-67A, 2 shts.)	8.00
_Waco YKS-6 Cabin	(#2-17A, 2 shts.)	8.00
_Waterman "Arrowbile"	(#3-25A, 2 shts.)	8.00
_Waterman "Gosling" Racer	(#1-4B, 2 shts.)	8.00
_Wright F2W-1 Racer	(#5-43A, 1 sht.)	4.50
_Wright F2W-2 Racer	(#5-44A, 2 shts.)	8.00
_Wright Brothers 1903 Flyer	(#0-123A, 2 shts.)	8.00
_Wright-Martin V	(#1-2A, 2 shts.)	8.00

send to: **Aviation Heritage**
P.O. Box 2065
Terre Haute, IN 47802